Introduction to Lens Design

Optical lenses have many important applications, from telescopes and spectacles, to microscopes and lasers. This concise, introductory book provides an overview of the subtle art of lens design. It covers the fundamental optical theory, and the practical methods and tools employed in lens design, in a succinct and accessible manner. Topics covered include first-order optics, optical aberrations, achromatic doublets, optical relays, lens tolerances, designing with off-the-shelf lenses, miniature lenses, and zoom lenses. Covering all the key concepts of lens design, and providing suggestions for further reading at the end of each chapter, this book is an essential resource for graduate students working in optics and photonics, as well as for engineers and technicians working in the optics and imaging industries.

JOSÉ SASIÁN is Professor of Optical Design at the James C. Wyant College of Optical Sciences at the University of Arizona in Tucson, AZ. He has taught a course on lens design for more than 20 years and has published extensively in the field. He has worked as a consultant in lens design for the optics industry, and has been responsible for the design of a variety of successful and novel lens systems.

Introduction to Lens Design

JOSÉ SASIÁN
University of Arizona

CAMBRIDGE
UNIVERSITY PRESS

University Printing House, Cambridge CB2 8BS, United Kingdom

One Liberty Plaza, 20th Floor, New York, NY 10006, USA

477 Williamstown Road, Port Melbourne, VIC 3207, Australia

314-321, 3rd Floor, Plot 3, Splendor Forum, Jasola District Centre, New Delhi - 110025, India

79 Anson Road, #06-04/06, Singapore 079906

Cambridge University Press is part of the University of Cambridge.

It furthers the University's mission by disseminating knowledge in the pursuit of education, learning and research at the highest international levels of excellence.

www.cambridge.org
Information on this title: www.cambridge.org/9781108494328
DOI: 10.1017/9781108625388

© José Sasián 2019

First published 2019

A catalogue record for this publication is available from the British Library

Library of Congress Cataloging in Publication data
Names: Sasián, José M., author.
Title: Introduction to lens design / José Sasián, University of Arizona.
Description: Cambridge, United Kingdom ; New York, NY, USA : University Printing House, 2019. | Includes bibliographical references and index.
Identifiers: LCCN 2019019484 | ISBN 9781108494328 (hardback)
Subjects: LCSH: Lenses–Design and construction.
Classification: LCC QC385.2.D47 S27 2019 | DDC 681/.423–dc23
LC record available at https://lccn.loc.gov/2019019484

ISBN 978-1-108-49432-8 Hardback

With appreciation to my lens design students.

This is evident even more when we realize that the combinations of lenses are very capricious entities, which in certain arrangements, probably because of laws deeply hidden in the building blocks of complicated functions, will give either not a good image at all, or one that is inevitably curved or distorted, and one understands easily that a lack of knowledge of these laws can lead to high costs and great useless efforts.

Joseph Maximillian Petzval
Bericht über die Ergebnisse einiger dioptrischen
Untersuchungen (Pest, 1843)

Contents

Preface

I have been fortunate to have taught for many years the course Lens Design OPTI517 at the James C. Wyant College of Optical Sciences at the University of Arizona. The thrust of this course is to help graduate students to acquire the skill of lens design and obtain a solid foundation in the subject in the space of about 16 weeks, which is the duration of the Fall academic term. Behind the scenes, the challenge has been how to completely and effectively achieve this thrust. This book is the result of teaching OPTI517 for about 20 years, and outlines the essential material interested students or optical engineers should know.

I have had the support and help from many individuals and I would like to acknowledge and to thank them. Robert Shannon handed me OPTI517, which he initiated and taught for many years at the then Optical Sciences Center. My colleagues Russell Chipman, John Greivenkamp, Angus Macleod, Jim Burge, Yuzuru Takashima, Tom Milster, Ron Liang, Buddy Martin, Hong Hua, Jim Schwiegerling, Roland Shack, Masud Mansuripur, Roger Angel, Stanley Pau, Bill Wolfe, Roy Frieden, Brian Anderson, Arvind Marathay, Rolf Binder, Dae Woo Kim, Eustace Dereniak, Steve Jacobs, Harry Barrett, Charles Falco, Jack Gaskill, John Koshel, and Dan Vukobratovich have been helpful and inspirational. I also would like to thank Richard Powell, Jim Wyant, and Tom Koch for the support they have provided me.

I have been fortunate to receive a "yes" when I asked many experts to visit the University of Arizona and help me teach lectures in optical design. Among the many individuals that I can recall and would like to acknowledge and thank are Richard Juergens, Bill Cassarly, Rich Pfisterer, Michael Humphreys, Akash Arora, Vini Mahajan, Richie Youngworth, Richard Buchroeder, Donald Dilworth, Mary Turner, Michael Gauvin, Craig Pansing, Dave Shafer, Jay Wilson, Jake Jacobsen, and John Rogers.

I would like to acknowledge and thank the lens design software companies, Lambda Research Corporation, Optenso™, Optical Systems Design, Inc.,

Synopsis®, and Zemax for always providing excellent academic access to their lens design software and for their outstanding support.

Richard Buchroeder, William Hicks, Craig Pansing, and Jim Schwiegerling provided many useful comments and suggestions to improve several chapters in this book. I am grateful for their help in this endeavor.

I would like thank Nicholas Gibbons, Sarah Lambert, and Roisin Munnelly, at Cambridge University Press, Vinithan Sethumadhavan at SPi-Global, and Liz Steel for their excellent editorial work in publishing this book.

1

Introduction

Lens design is an exciting and important field of optics. This field provides designs for a great diversity of lens and mirror systems needed in many other fields, such as consumer optics, microscope optics, telescope optics, lenses for optical lithography, and photographic optics. Lens and mirror systems are ubiquitous. The work of a lens designer is to provide the constructional data and fabrication tolerances of all the optical elements that a given lens system requires to perform the intended function. Currently many students and engineers are interested in lens design because the field by itself is of great interest, or because they have the need to analyze and design lens systems required in their engineering practice. An optical engineer should have at least some familiarity with how a lens system is designed so that he or she can effectively contribute to develop optical systems.

1.1 Aims of This Book

This book is an introduction to lens design, and has been written to provide an overview of topics that are indispensable to acquire the skill of lens design. Acquiring this skill, the skill of lens design, requires learning some theory, learning how to use lens design software, and gaining experience by designing actual lenses. This book will help the interested reader to understand the theory and methods used in lens design. The book does not have lengthy discussions but, rather, brief discussions to point out essential knowledge. A few references are given for further reading, where the reader can deepen his or her knowledge about a topic.

There are many excellent books about lens design, such as *Lens Design Fundamentals* by Rudolf Kingslake and Barry Johnson, and *Modern Lens Design* by Warren Smith. However, these and other comprehensive books might not be appropriate for an introduction to lens design, as part of their main focus is the design and survey of a variety of specific lenses. Instead, this introductory book intends to be brief and also to give an overview of topics that a current optical engineer needs to know about lens system design. A graduate student or an optical engineer who understands the content of this book and models, in a lens design program, the different lens systems discussed in it, would then have a solid foundation to practice the skill of lens design. Another aim of this book is to provide an efficient introduction to lens design to an interested student or optical engineer, so that he or she is well positioned to analyze, combine, debug, adjust, or design lens systems.

1.2 Topics Covered

Essential to lens design, and to optical engineering, is an understanding about how optical aberrations are corrected, balanced, or minimized. The reader should have some familiarity with first-order optics and with the theory of optical aberrations, as many discussions revolve about the choices made in the layout of a lens system and how to correct the aberrations. In this book, structural aberration coefficients are used to determine primary aberrations and to understand how to correct, balance, or minimize them. Chapter 2 provides a review of first-order optics and aberrations. Chapter 3 provides a brief discussion of aspheric surfaces. Chapter 4 provides a discussion of thin lenses and how aberrations are controlled in very simple lens systems. Chapter 5 provides a discussion about how ray tracing takes place, and some useful techniques. Chapter 6 provides a discussion about radiometric aspects of a lens system, which are important for a more comprehensive understanding of how lenses work. Chapter 7 discusses achromatic and athermal lenses. Chapter 8 provides a number of lens examples that use combinations of achromatic doublets. This chapter is insightful because it shows how lenses are combined. Chapter 9 discusses the tools used to determine image quality. Chapter 10 discusses how to perform a tolerancing analysis for providing tolerances to the constructional parameters of a lens system that will be manufactured. Chapter 11 comments on issues in using a lens design program. Chapter 12 discusses three classical lenses; the Petzval portrait objective, the Cooke triplet, and the double Gauss lens. Chapter 13 discusses issues that arise in combining lens systems; it also contributes to providing a more comprehensive view about lens systems. Chapter 14 discusses designing with off-the-shelf lenses. Chapter 15 discusses ghost images in a lens system. Chapter 16 discusses some basic mirror systems. Chapter 17 discusses miniature lenses. Chapter 18 provides basic concepts in zoom lens design. In addition to a glossary of terms, the book provides five Appendices, where several tables related to aberrations are provided, as well as a discussion of the sine condition. Thus, the book contents provide a shift in the way lens design is taught. In this introductory book there is more emphasis on providing a broader view of fundamentals and essential topics in lens design, rather than bringing attention to the detailed design of a survey of lenses. This shift responds to current needs in the optical industry, and modern approaches to learning. Yet, this book provides a solid introduction for those who would like to specialize in the art of lens design.

1.3 The Art of Lens Design

There are many types of lens systems, and their variety is increasing with advancements and the creation of new technological fields. Examples of lens types are projection lenses, telephoto lenses, convertible lenses, catadioptric lenses, zoom lenses, underwater lenses, lenses for aerial photography, anamorphic lenses, panoramic lenses, lenses for video and cinematography, lenses for scanning, relay lenses, periscope lenses, and lenses for endoscopes.

The process of lens design starts with understanding the application the intended lens is to be designed for. From understanding the application, the lens specifications list follows. This list of specifications is not always complete or correct. A lens designer must make efforts to verify that the specifications list is as complete and correct as possible. The lens specifications may involve first-order, packaging, image quality, environmental, and lens fabrication constraints and requirements. Once the specifications are understood, the lens designer may start a design from first principles, and by adding complexity to simple lenses. A first-order lens layout can help to visualize a given lens and determine, for example, lens size, element optical power, and type of lens configuration. From the first-order layout, considerations are made about how the aberrations could be corrected. Then a lens design program is used to model and optimize the lens system, and to find alternative lens solutions for comparison. A lens analysis is also made to determine tolerances that a lens manufacturer would need to make the lens elements. A lens design can also start from existing lenses in the patent literature. A lens designer should have effective communication with the opto-mechanical engineer and lens manufacturer to make sure that the designed lens can be mounted in a barrel, fabricated, and assembled. Lens drawings are then drafted. Some optical engineers may not actually design lenses, but would analyze, debug, adjust, and combine existing lens systems. A critical design review is often held to approve, or disapprove, a lens for fabrication. The overall process of lens design is also of exercising design creativity, and this in part is what makes lens design an exciting field.

Further Reading

Bentley, Julie L., Olson, Craig, Youngworth, Richard N. "In the era of global optimization, the understanding of aberrations remains the key to designing superior optical systems," Proceedings of SPIE 7849, Optical Design and Testing IV, 78490C (2010); doi: 10.1117/12.871720.

Kidger, M. J. "The importance of aberration theory in understanding lens design," *Proceedings of SPIE*, 3190 (1997), 26–33.

Sasián, J. "Trends in teaching lens design," *Proceedings of SPIE*, 4588 (2001), 56–58.

Sasián, José. "From the landscape lens to the planar lens: a reflection on teaching lens design," Proceedings of SPIE 5865, Tribute to Warren Smith: A Legacy in Lens Design and Optical Engineering, 58650I (2005); doi: 10.1117/12.624566.

Shannon, Robert R. "Teaching of lens design," Proceedings of SPIE 1603, Education in Optics (1992); https://doi.org/10.1117/12.57848.

2

Classical Imaging, First-Order Imaging, and Imaging Aberrations

This chapter provides a brief overview of essential imaging concepts used in lens design. Whether classical imaging, which is congruent with first-order optics, is required in a lens system, or any other type of imaging, depends on system application. Therefore, a clear understanding of what imaging is and of departures from such imaging, called aberrations, is essential for a lens design practice.

2.1 Classical Imaging

The main goal in lens design is the design of imaging lenses where images, particularly sharp, are formed. Then it is important to discuss the concept of an image. Depending on application, different imaging concepts can be devised. However, classical imaging, where the image is a scaled copy of the object, is often required for a lens system. The underlying mechanism for classical imaging is central projection. Object points are projected into image points on an image plane, by the line defined by an object point on the object plane and a central projection point pair as shown in Figure 2.1. The projection point pair is the center of perspective and in a lens system, which we assume to have axial symmetry, is represented by a nodal point in object space and its conjugate point, the nodal point in image space. The main attributes of a classical image are its location and its size. The Newtonian or Gaussian imaging equations shown in Table 2.1 permit calculating these attributes and represent central projection imaging.

Ideal imaging as defined by central projection is often a designing goal. For an object at infinity that subtends a semi-field of view, θ, the image height, $\bar{y}_i$, measured from the optical axis, is related to the focal length, f, by the mapping, $\bar{y}_i = f \cdot \tan(\theta)$. However, according to application, there are other

Table 2.1 *Imaging equations*

Newtonian equations	Gaussian equations
$\frac{z}{f} = -\frac{1}{m}$	$\frac{f'}{z'} + \frac{f}{z} = 1$
$\frac{z'}{f'} = -m$	$\frac{z}{f} = 1 - \frac{1}{m}$
$zz' = ff'$	$\frac{z'}{f'} = 1 - m$
The object and image distances z and z' are measured, respectively, from the front and rear focal points. f and f' are the front and rear focal lengths.	The object and image distances z and z' are measured, respectively, from the front and rear principal points. The transverse magnification is m.

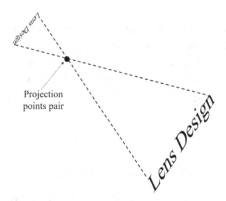

Figure 2.1 Central projection imaging where the object on the left is imaged on the right. In this case the projection points coincide in space.

possible mappings such as the equidistant mapping, $\bar{y}_i = f \cdot \theta$, or the orthographic mapping, $\bar{y}_i = f \cdot \sin(\theta)$. We are assuming that the object and image lay on planes perpendicular to the optical axis of the optical system. There are some applications that require the image to lay on a curved surface, and then the concept of classical imaging no longer applies.

The point is that lenses are designed to produce images which require a lens designer to be clear about what imaging means. Imaging is and will continue to be an important subject which substantially impacts lens design. What imaging is depends on system application.

2.2 First-Order Optics

The concept of first-order imaging arises from a first-order approximation to the path of a real ray. A real ray in homogenous media travels in straight lines, refracts according to Snell's law, $n' \sin(I') = n \sin(I)$, and its tracing considers the actual shape of the refracting surface. A first-order ray refracts according to a first-order approximation to Snell's law, $n'i' = ni$, and treats the optical surfaces as planar, but with refracting power, ϕ. To trace a first-order ray, the refraction and transfer equations are used:

$$n'u' = nu - y\phi \tag{2.1}$$

$$y' = y + u't, \tag{2.2}$$

where u and u' are the slopes of the ray before and after refraction, y is the ray height at the surface which is assumed planar but with optical power ϕ, n is the index of refraction, and t is the distance to the next surface.

In lens design we are concerned with first-order imaging, as obtained by tracing first-order rays, because it is equivalent to central projection imaging. In addition, first-order imaging establishes a model for a lens system where the cardinal points – these are the focal points, the nodal points, and the principal points – have specific ray properties and serve as useful references. Many calculations in lens design are done by tracing first-order rays and, therefore, an optical designer must be familiar with first-order optics. An example of a calculation in lens design software is what is known as a "solve" in which the program automatically sets the distance t from the last surface to the ideal image plane using $t = -y/u'$.

The space where the object resides is called the object space and is infinite in extent. Similarly, the space where the image resides is called the image space and is infinite in extent. An important structure in a lens system is the aperture stop. The aperture stop is assumed to be circular, to lay on a plane perpendicular to the optical axis, and it solely limits the amount of light for the on-axis beam. The aperture stop helps to well define a lens system; this is, light beams for every field point become well defined after they pass through the aperture stop. The image of the aperture stop in object space is defined as the entrance pupil, and the image of the aperture stop in image space is defined as the exit pupil. The pupils and the stop are optically conjugated, meaning that their locations and sizes satisfy the Newtonian or Gaussian equations that are summarized in Table 2.1. Another aperture that contributes to well define a lens system is the field stop. The field stop limits the field of view of a lens system, and ideally it is located at an image plane.

Rays that travel in a plane that contains the lens system axis of rotational symmetry are called meridional rays. Rays that do not travel in a meridional plane are called skew rays. Two important first-order rays are the marginal and chief rays. By definition, the marginal ray is a meridional ray that originates at the on-axis object point and passes through the edge of the aperture stop. The chief ray is a meridional ray that originates at the edge of the field of view and passes through

Table 2.2 *First-order concepts*

Optical axis	The axis about which an optical system has rotational symmetry.
Object space	The space where the object resides, which is assumed infinite in extent.
Image space	The space where the image resides, which is assumed infinite in extent.
Aperture stop	The aperture that solely limits the amount of light for the axial light beam.
f	Front focal length.
f'	Rear focal length.
Optical power or Refractive power (ϕ)	$\phi = -\frac{n}{f} = \frac{n'}{f'}$; n is the index of refraction in object space, and n' is the index in image space. The unit of power is the diopter or 1/meter.
Effective focal length (EFL)	The inverse of the optical power.
$F/\#$, F-number	The effective focal length divided by the diameter of the entrance pupil. $F/\# = \frac{EFL}{2y_e}$
Lagrange invariant ($\mathcal{K}$)	It relates to the optical throughput or capacity of an optical system to transfer optical power. $\mathcal{K} = n\bar{u}y - nu\bar{y}$
Afocal	The focal lengths are not defined.
Telecentricity in object space	The image of the aperture stop in object space is at infinity. Equivalently, the chief ray in object space is parallel to the optical axis.
Telecentricity in image space	The image of the aperture stop in image space is at infinity. Equivalently, the chief ray in image space is parallel to the optical axis.
Transverse magnification (m)	The first-order ratio of the image size to the object size.

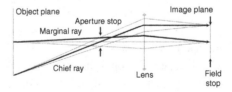

Figure 2.2 The marginal and chief rays (highlighted in bold) in relation to the aperture stop, the object and image planes, the field stop, and an ideal lens.

the center of the aperture stop. The trace of these two rays permits obtaining useful information about the imaging of an optical system. Figure 2.2 shows an object plane, an aperture stop, a lens, an image plane, and two sets of rays defining two light beams for the on-axis object point and for an off-axis point. In particular, Figure 2.2 illustrates the marginal and chief rays using bold rays. Table 2.2 provides a glossary of first-order concepts, and Table 2.3 provides a summary of first-order quantities. The Lagrange invariant, $\mathcal{K}$, is defined by

Table 2.3 *Marginal and chief first-order rays' related quantities*

Item	Marginal ray	Chief ray
Object/pupil distance	s	$\bar{s}$
Image/pupil distance	s'	$\bar{s}'$
Ray slope of incidence	$i = u - \alpha$	$\bar{i} = \bar{u} - \bar{\alpha}$
Ray height at surface	y	$\bar{y}$
	y_e	$\bar{y}_o$
	y_s	$\bar{y}_i$
Ray slope	$u = -y/s$	$\bar{u} = -\bar{y}/\bar{s}$
Normal line slope	$\alpha = -y/r = u - i$	$\bar{\alpha} = -\bar{y}/r = \bar{u} - \bar{i}$
Refraction invariant	$A = ni = n\left(\frac{1}{r} - \frac{1}{s}\right)y$	$\bar{A} = n\bar{i} = n\left(\frac{1}{r} - \frac{1}{\bar{s}}\right)\bar{y}$
Surface radius	r	
Surface vertex curvature	c	
Thickness to next surface	t	
Surface optical power	$\phi = \frac{n'-n}{r}$	
Lagrange invariant	$\mathcal{K} = n\bar{u}y - nu\bar{y} = \bar{A}y - A\bar{y}$	

Quantities related to the chief ray carry a bar.
Primed quantities refer to the image space and un-primed to the object space.

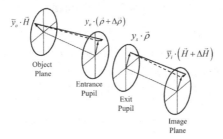

Figure 2.3 Model of an axially symmetric optical system showing the path of a first-order ray and the path of a real ray.

$\mathcal{K} = n\bar{u}y - nu\bar{y}$, using the slope and height of the marginal and chief rays. Its value does not depend on the transverse plane where it is calculated. The amount of optical flux, or optical throughput, $T = \pi^2 \mathcal{K}^2$, that can pass through an optical system is proportional to the square of the Lagrange invariant.

Figure 2.3 provides a representation of an optical system where the object and image planes and the entrance and exit pupils are shown. The solid line represents a real ray traveling from the object plane to the image plane, and the broken line represents a first-order ray. Two points are required to define a ray; the first point is defined by the field vector, $\vec{H}$, which lies in the object plane, and the second point is defined by the aperture vector, $\vec{\rho}$, which lies in the exit pupil plane. Both vectors are normalized so their magnitudes range from 0 to 1. To indicate an actual field point, the field vector is scaled by the chief ray height in object space, $\bar{y}_o$, and the aperture vector is scaled by the marginal ray height, y_s, at the exit pupil. In Figure 2.3 the real ray and the first-order ray coincide necessarily, per definition, at the object plane and at the exit pupil plane. Everywhere else these rays may differ in path. In particular, at the image plane they differ by the vector $\bar{y}_i \Delta \vec{H}$, and at the entrance pupil plane by the vector $y_e \Delta \vec{p}$. They differ at the image plane because of image defects known as image aberrations; similarly they differ at the entrance pupil because of pupil aberrations.

A lens designer may start a design with a first-order lens layout, as shown in Figure 2.4. Two ideal lenses with the same focal length form a $4f$ relay, as the distance between object and image is four-times the focal length of the lens elements. Such a layout provides useful information such as the ideal path of rays, the diameter of the lens elements, and the system's size.

In sum, first-order optics is equivalent to classical imaging and provides a basic structure to model a lens system.

2.3 Imaging Aberrations

Actual lens systems do not produce perfect imaging, but introduce image defects known as optical aberrations. Aberration can refer to wave aberration, transverse, longitudinal, or angular ray aberration.

In relation to Figure 2.5 the Optical Path Length (*OPL*) along a ray is defined as,

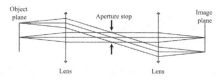

Figure 2.4 A doubly telecentric relay system in a first-order layout. Rays from on-axis and off-axis field points are shown. Two ideal positive lenses are schematically drawn as vertical lines with arrow ends.

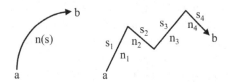

Figure 2.5 Left: Path of a ray in an inhomogeneous medium. Right: Path of a ray in several homogenous media.

$$OPL = \int_a^b n(s)ds, \tag{2.3}$$

where $n(x, y, z)$ is the index of refraction as a function of position, and ds is the element of arc length. If the index of refraction is uniform from medium to medium, then the *OPL* reduces to a summation over the different media,

$$OPL = \sum_i n_i s_i, \tag{2.4}$$

where n_i is the index of refraction, and s_i is the ray length in medium i. The units of *OPL* are of length, for example, millimeters. If the *OPL* is divided by the speed of light, then we obtain a transit time from point a to point b along the ray.

Taking an object point as the origin of rays, the geometrical wavefront is defined as the locus of constant optical path length. As shown in Figure 2.6, in a homogenous medium the wavefront is spherical in shape. However, when the wavefront propagates through an optical system, it is deformed, and its shape is no longer spherical. As the rays are normal to the wavefront, they no longer converge to a sharp image point; i.e., the ideal image point as defined by central projection. In relation to Figure 2.7, the wavefront aberration represents

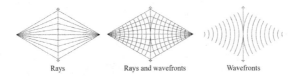

Rays   Rays and wavefronts   Wavefronts

Figure 2.6 Rays and waves which diverge from a point source are refracted by a lens and converge to a point image.

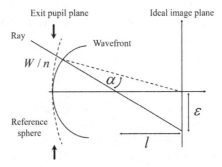

Figure 2.7 Different metrics to refer to aberrations as wavefront deformation, angular ray aberration, transverse ray aberration, or longitudinal ray aberration.

wave deformation, W/n, and the ray error represents angular, α, transverse, ε, or longitudinal, l, aberration.

The wavefront deformation is measured with respect to a reference sphere. As shown in Figure 2.8, the reference sphere is centered at the ideal image point and passes by the on-axis exit pupil point. Note that, because of transverse ray aberration, $\vec{\varepsilon} = \bar{y}_i \Delta \vec{H}$, the ray does not intersect the ideal image point at $\bar{y}_i \vec{H}$. Given a ray defined by the field and aperture vectors $\vec{H}$ and $\vec{\rho}$, the distance along the ray between the reference sphere and the actual wavefront times the index of refraction in image space is the wavefront deformation from the reference sphere for that ray.

For an axially symmetric system the aberration function, $W\left(\vec{H}, \vec{\rho}\right)$, provides the geometrical wavefront deformation at the exit pupil as a function of the normalized field, $\vec{H}$, and aperture, $\vec{\rho}$, vectors. The field vector is located at the object plane and defines where a given ray originates from. The aperture vector defines the intersection of a given ray with the pupil plane. The aperture vector is usually located at the exit pupil plane, but it can also be located at the entrance pupil plane. Figure 2.9 shows in image space the ideal image of the field vector and the aperture vector at the exit pupil plane. The aberration function, being a scalar, involves dot products of the field and aperture vectors, specifically $\vec{H} \cdot \vec{H}, \vec{H} \cdot \vec{\rho}$, and $\vec{\rho} \cdot \vec{\rho}$. These dot products only

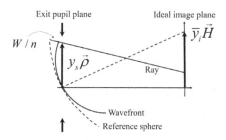

Figure 2.8 The wavefront deformation W/n is determined with the aid of a reference sphere.

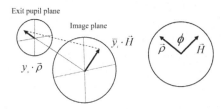

Figure 2.9 The field and aperture vectors (scaled by the marginal ray height at the exit pupil and the chief ray height at the image plane) and the angle between them, looking down the optical axis.

depend on the magnitude of the vectors and on the cosine of the angle, ϕ, between them. The dot products are used to describe axial symmetry, and are known as the rotational invariants, since they do not change their magnitude upon a rotation of the coordinate system about the optical axis.

The aberration function provides the wavefront deformation in terms of optical path, as measured along a particular ray (defined by the tip of the field vector and the tip of the aperture vector) and from the reference sphere to the wavefront. Equivalently, the aberration function provides the Optical Path Difference (OPD) between the OPL from the object to the wavefront at the exit pupil, and the OPL from the object point to the reference sphere. The aberration function is written to sixth-order of approximation as:

$$
\begin{aligned}
W\left(\vec{H},\vec{\rho}\right) = \sum_{j,m,n} & W_{k,l,m}\left(\vec{H}\cdot\vec{H}\right)^{j}\left(\vec{H}\cdot\vec{\rho}\right)^{m}\left(\vec{\rho}\cdot\vec{\rho}\right)^{n} \\
= & W_{000} + W_{200}\left(\vec{H}\cdot\vec{H}\right) + W_{111}\left(\vec{H}\cdot\vec{\rho}\right) + W_{020}\left(\vec{\rho}\cdot\vec{\rho}\right) \\
& + W_{040}\left(\vec{\rho}\cdot\vec{\rho}\right)^{2} + W_{131}\left(\vec{H}\cdot\vec{\rho}\right)\left(\vec{\rho}\cdot\vec{\rho}\right) + W_{222}\left(\vec{H}\cdot\vec{\rho}\right)^{2} \\
& + W_{220}\left(\vec{H}\cdot\vec{H}\right)\left(\vec{\rho}\cdot\vec{\rho}\right) + W_{311}\left(\vec{H}\cdot\vec{H}\right)\left(\vec{H}\cdot\vec{\rho}\right) + W_{400}\left(\vec{H}\cdot\vec{H}\right)^{2} \\
& + W_{240}\left(\vec{H}\cdot\vec{H}\right)\left(\vec{\rho}\cdot\vec{\rho}\right)^{2} + W_{331}\left(\vec{H}\cdot\vec{H}\right)\left(\vec{H}\cdot\vec{\rho}\right)\left(\vec{\rho}\cdot\vec{\rho}\right) \\
& + W_{422}\left(\vec{H}\cdot\vec{H}\right)\left(\vec{H}\cdot\vec{\rho}\right)^{2} + W_{420}\left(\vec{H}\cdot\vec{H}\right)^{2}\left(\vec{\rho}\cdot\vec{\rho}\right) \\
& + W_{511}\left(\vec{H}\cdot\vec{H}\right)^{2}\left(\vec{H}\cdot\vec{\rho}\right) + W_{600}\left(\vec{H}\cdot\vec{H}\right)^{3} + W_{060}\left(\vec{\rho}\cdot\vec{\rho}\right)^{3} \\
& + W_{151}\left(\vec{H}\cdot\vec{\rho}\right)\left(\vec{\rho}\cdot\vec{\rho}\right)^{2} + W_{242}\left(\vec{H}\cdot\vec{\rho}\right)^{2}\left(\vec{\rho}\cdot\vec{\rho}\right) + W_{333}\left(\vec{H}\cdot\vec{\rho}\right)^{3}
\end{aligned}
$$

$$(2.5)$$

where the sub-indices, j, m, n, represent integers, and $k = 2j + m$, $l = 2n + m$, and $W_{k,l,m}$ represent aberration coefficients. The terms in the aberration function represent aberrations, that is, basic forms in which the wavefront can be deformed. The sum of all aberration terms and orders produces the actual total wavefront deformation. The order of an aberration term is given by $2\cdot(j + m + n)$, which is always an even order. In the aberration function the field and aperture vectors are normalized so that, when they are unity, the coefficients represent the maximum amplitude of each aberration, which is expressed in wavelengths. The lower indices k, l, m in each coefficient indicate, respectively, the algebraic power of the field vector, the aperture vector, and the cosine of the angle ϕ between these vectors.

Table 2.4 summarizes the first four orders of aberrations using both vector and algebraic expressions. The fourth-order terms are often called the primary aberrations. The ten sixth-order terms can be divided into two groups. The first group (first six terms) can be considered as an improvement upon the primary

Table 2.4 *Wavefront aberrations*

Aberration name	Vector form	Algebraic form	j	m	n
Zero-order					
Uniform piston	W_{000}	W_{000}	0	0	0
Second-order					
Quadratic piston	$W_{200}\left(\vec{H}\cdot\vec{H}\right)$	$W_{200}H^2$	1	0	0
Magnification	$W_{111}\left(\vec{H}\cdot\vec{\rho}\right)$	$W_{111}H\rho\cos(\phi)$	0	1	0
Focus	$W_{020}\left(\vec{\rho}\cdot\vec{\rho}\right)$	$W_{020}\rho^2$	0	0	1
Fourth-order					
Spherical aberration	$W_{040}\left(\vec{\rho}\cdot\vec{\rho}\right)^2$	$W_{040}\rho^4$	0	0	2
Coma	$W_{131}\left(\vec{H}\cdot\vec{\rho}\right)\left(\vec{\rho}\cdot\vec{\rho}\right)$	$W_{131}H\rho^3\cos(\phi)$	0	1	1
Astigmatism	$W_{222}\left(\vec{H}\cdot\vec{\rho}\right)^2$	$W_{222}H^2\rho^2\cos^2(\phi)$	0	2	0
Field curvature	$W_{220}\left(\vec{H}\cdot\vec{H}\right)\left(\vec{\rho}\cdot\vec{\rho}\right)$	$W_{220}H^2\rho^2$	1	0	1
Distortion	$W_{311}\left(\vec{H}\cdot\vec{H}\right)\left(\vec{H}\cdot\vec{\rho}\right)$	$W_{311}H^3\rho\cos(\phi)$	1	1	0
Quartic piston	$W_{400}\left(\vec{H}\cdot\vec{H}\right)^2$	$W_{400}H^4$	2	0	0
Sixth-order					
Oblique spherical aberration	$W_{240}\left(\vec{H}\cdot\vec{H}\right)\left(\vec{\rho}\cdot\vec{\rho}\right)^2$	$W_{240}H^2\rho^4$	1	0	2
Coma	$W_{331}\left(\vec{H}\cdot\vec{H}\right)\left(\vec{H}\cdot\vec{\rho}\right)\left(\vec{\rho}\cdot\vec{\rho}\right)$	$W_{331}H^3\rho^3\cos(\phi)$	1	1	1
Astigmatism	$W_{422}\left(\vec{H}\cdot\vec{H}\right)\left(\vec{H}\cdot\vec{\rho}\right)^2$	$W_{422}H^4\rho^2\cos^2(\phi)$	1	2	0
Field curvature	$W_{420}\left(\vec{H}\cdot\vec{H}\right)^2\left(\vec{\rho}\cdot\vec{\rho}\right)$	$W_{420}H^4\rho^2$	2	0	1
Distortion	$W_{511}\left(\vec{H}\cdot\vec{H}\right)^2\left(\vec{H}\cdot\vec{\rho}\right)$	$W_{511}H^5\rho\cos(\phi)$	2	1	0
Piston	$W_{600}\left(\vec{H}\cdot\vec{H}\right)^3$	$W_{600}H^6$	3	0	0
Spherical aberration	$W_{060}\left(\vec{\rho}\cdot\vec{\rho}\right)^3$	$W_{060}\rho^6$	0	0	3
	$W_{151}\left(\vec{H}\cdot\vec{\rho}\right)\left(\vec{\rho}\cdot\vec{\rho}\right)^2$	$W_{151}H\rho^5\cos(\phi)$	0	1	2
	$W_{242}\left(\vec{H}\cdot\vec{\rho}\right)^2\left(\vec{\rho}\cdot\vec{\rho}\right)$	$W_{242}H^2\rho^4\cos^2(\phi)$	0	2	1
	$W_{333}\left(\vec{H}\cdot\vec{\rho}\right)^3$	$W_{333}H^3\rho^3\cos^3(\phi)$	0	3	0

aberrations by their increased field dependence, and the second group (last four terms) represents new wavefront deformation forms. Figure 2.10 shows the shape (aperture dependence only) of the zero, second, fourth, and the new wavefront shapes of the sixth-order aberrations.

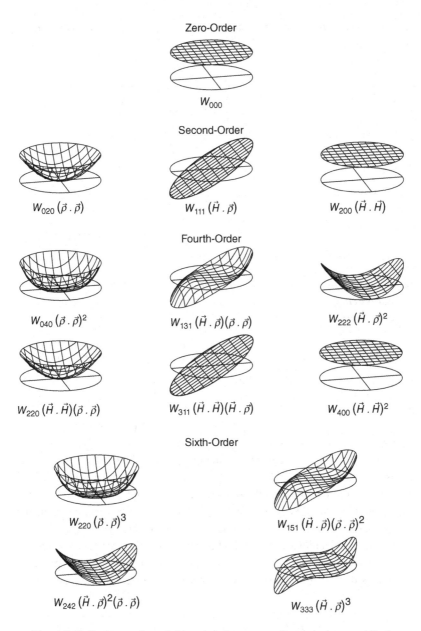

Figure 2.10 Basic wavefront deformation shapes according to symmetry. All of the basic deformations are either axially symmetric, double plane symmetric, or plane symmetric.

(Grid figures with permission and courtesy of Roland Shack.)

Table 2.5 *Aberration coefficients for a system of* j *spherical surfaces in terms of Seidel sums*

Coefficient	Seidel sum
$W_{040} = \frac{1}{8}S_I$	$S_I = -\sum_{i=1}^{j} \left(A^2 y \Delta\left(\frac{u}{n}\right) \right)_i$
$W_{131} = \frac{1}{2}S_{II}$	$S_{II} = -\sum_{i=1}^{j} \left(A\bar{A} y \Delta\left(\frac{u}{n}\right) \right)_i$
$W_{222} = \frac{1}{2}S_{III}$	$S_{III} = -\sum_{i=1}^{j} \left(\bar{A}^2 y \Delta\left(\frac{u}{n}\right) \right)_i$
$W_{220} = \frac{1}{4}(S_{III} + S_{IV})$	$S_{IV} = -\mathcal{K}^2 \sum_{i=1}^{j} P_i$
$W_{311} = \frac{1}{2}S_V$	$S_V = -\sum_{i=1}^{j} \left(\bar{A}\left[\bar{A}^2 \Delta\left(\frac{1}{n^2}\right)y - P(\mathcal{K} + \bar{A}y)\bar{y} \right] \right)_i$

Table 2.6 *Quantities derived from first-order ray data used in computing the aberration coefficients*

Refraction invariant marginal ray	Refraction invariant chief ray	Lagrange invariant	Surface curvature	Petzval sum term
$A = ni = nu + nyc$	$\bar{A} = n\bar{i} = n\bar{u} + n\bar{y}c$	$\mathcal{K} = n\bar{u}y - nu\bar{y}$ $= \bar{A}y - A\bar{y}$	$c = \frac{1}{r}$	$P = c \cdot \Delta\left(\frac{1}{n}\right)$

In Table 2.4, the piston terms represent a uniform phase change across the aperture that does not degrade the image quality. Physically, piston terms represent a time delay or advance in the time of arrival of the wavefront as it propagates from the object to the exit pupil. The second-order term magnification represents a change of magnification, and the focus term represents a change in the axial location of the image. The coefficients for magnification and focus are set to zero given that Gaussian and Newtonian optics accurately predict the size and location of an image. However, a focus term can be added to minimize aberrations or to focus light on a plane other than the ideal image plane.

2.4 Computing Aberration Coefficients

For an optical system made out of j spherical surfaces, the fourth-order aberration coefficients are determined by computing the Seidel sums, S_I, S_{II}, S_{III}, S_{IV}, and S_V. These sums depend only on the Lagrange invariant, and on quantities from a first-order marginal and chief ray trace. Table 2.5 provides

Table 2.7 *Ranges of field of view and relative apertures in lens systems*

Very small	Small	Medium	Large	Very large
FOV				
$< \pm\frac{1}{2}°$	$\pm\frac{1}{2}°$ to $5°$	$\pm5°$ to $25°$	$\pm25°$ to $45°$	$> \pm45°$
F/#				
< 1	1 to 4	4 to 8	8 to 16	> 16

the Seidel sums formulae, and Table 2.6 provides first-order quantities used in the computation.

In Table 2.5 the operator, $\Delta()$, gives the difference of the argument after and before refraction, this is $\Delta(n) = n' - n$. The summation symbol indicates that the total amount of aberration, for example for spherical aberration W_{040}, is the sum of the spherical aberration contributed by each spherical surface in the system.

2.5 Field of View and Relative Aperture

Two important specifications for a lens system are its field of view (FOV) and its relative aperture (F/#). The field of view is the observable scene of a lens system for which it is designed. For a lens that works at finite conjugates, the FOV is specified by the object or image size, and giving the height, width, or both. When the object is at infinity the field of view is usually specified by the semi-angle subtended by the scene, or object, as seen from the entrance pupil, either horizontally, vertically, or both. It can also be specified by the height, width, or both, of the image.

The relative aperture is defined as the ratio of the effective focal length EFL to the diameter of the entrance pupil D_E. Also known as $F/\#$, FNO, F-number, and focal ratio F. For lens systems that work at finite conjugates the effective relative aperture, sometimes referred to as the working $F/\#$, is given by $F/\# = (1 - m)EFL/D_E$, where m is the transverse magnification.

In a lens that is free from spherical aberration the numerical aperture (NA) is defined by $NA = n \sin(\theta)$, where n is the index of refraction and θ is the angle of the real marginal ray with the optical axis. The NA can refer to the object or image spaces. In an aplanatic system the on-axis image brightness is proportional to the square of the numerical aperture, NA^2.

Being aware of the field of view and relative aperture in a lens system is important. Usually, the larger the field of view or the lower the F/# is, the more difficult it is to design a lens. Table 2.7 provides ranges of these specifications in lens systems.

Table 2.8 *Doublet lens optical specifications,* $\lambda = 0.6328$ µm

Focal length	F-number	Field of view	Aperture stop	Object location	Image quality
100 mm	5	±5°	At doublet	At infinity	Aplanatic

Table 2.9 *Doublet lens prescription*

Surface	Radius	Thickness	Glass
Stop	71.262	4	LAK33
2	−40.363	3	SF6
3	−1,237.921	96.028	Air
Image			Air

Units are millimeters.

Figure 2.11 First-order layout of the doublet lens. Two light beams are shown for the on-axis field point and for the 5 degrees off-axis field point.

Figure 2.12 Doublet lens drawn with lens surfaces.

2.6 Lens Design Example

A lens designer often starts with the lens specifications. Some important first-order lens specifications are the focal length, the F-number, and the field of view. The lens in consideration is a cemented doublet lens corrected for spherical aberration and coma, this is aplanatic. The specifications are given in Table 2.8, and a first-order layout is shown in Figure 2.11.

After a first-order lens layout is created to visualize the lens, a lens designer may substitute ideal lenses for real lenses and correct, balance, or minimize some aberrations. The doublet design is shown in Figure 2.12, where computer optimization was performed to provide a focal length of 100 mm and to correct for spherical aberration W_{040} and coma aberration W_{131}. There are three lens surfaces, and their curvatures are effective variables to satisfy the focal length requirement and the aberration correction.

Table 2.9 provides the doublet lens prescription, Table 2.10 provides a first-order ray trace, and Table 2.11 provides the Seidel sum calculation surface by surface.

Table 2.10 *First-order ray trace*

Surface	y	$n'u'$	$n'i'$	$\bar{y}$	$n'\bar{u}'$	$n'\bar{i}'$
1	10.00	−0.06	0.14	0.00	0.05	0.09
2	9.76	−0.05	−0.53	0.20	0.05	0.08
3	9.60	−0.10	−0.11	0.35	0.09	0.09

Table 2.11 *Doublet lens wave aberration coefficients,* $\lambda = 0.6328$ μm

Surface	W_{040}	W_{131}	W_{222}	W_{220}	W_{311}
1	1.34	3.33	2.08	2.86	3.57
2	−2.90	1.73	−0.26	−0.24	0.07
3	1.56	−5.06	4.10	2.16	−3.49
Total	0.00	0.00	5.92	4.77	0.14

Examination of Table 2.11 shows that surface curvatures were chosen to correct for spherical aberration and coma aberration. There are 5.92 waves of astigmatism aberration and 4.77 waves of field curvature aberration. Distortion aberration is negligible.

2.7 Stop Shifting

Stop shifting is the change of position along the optical axis of the aperture stop to a new location while maintaining the optical throughput, $T = \pi^2 \mathcal{K}^2$, of the system. This requires maintaining the $F/\#$ and, consequently, the aperture stop must change size. The parameter $\bar{S}$ quantifies stop shifting and can be computed at any surface of the optical system using the old and new quantities at that surface, as indicated by

$$\bar{S} = \frac{\bar{u}_{new} - \bar{u}_{old}}{u} = \frac{\bar{y}_{new} - \bar{y}_{old}}{y} = \frac{\bar{A}_{new} - \bar{A}_{old}}{A}, \qquad (2.6)$$

where $\bar{A} = n\bar{i}$ is the refraction invariant for the chief ray, $A = ni$ is the refraction invariant for the marginal ray, $\bar{u}$ is the chief ray slope, u is the marginal ray slope, $\bar{y}$ is the chief ray height at the surface, and y is the marginal ray height at the surface.

A useful set of formulas to determine the change of Seidel sum when the stop aperture is shifted along the optical axis is presented in Table 2.12, where the asterisk indicates the new value for the Seidel sum, and where $\bar{S}$ is the stop shifting parameter.

Table 2.12 *Seidel sums upon stop shifting*

$$S_I^* = S_I$$
$$S_{II}^* = S_{II} + \bar{S}S_I$$
$$S_{III}^* = S_{III} + 2{\cdot}\bar{S}S_{II} + \bar{S}^2 S_I$$
$$S_{IV}^* = S_{IV}$$
$$S_V^* = S_V + \bar{S}(S_{IV} + 3{\cdot}S_{III}) + 3{\cdot}\bar{S}^2 S_{II} + \bar{S}^3 S_I$$

Stop shifting formulas provide insight into how aberrations change upon stop shifting whenever there is aberration present in a system. For example, in the presence of spherical aberration, the amount of coma aberration can be changed by stop shifting according to $S_{II}^* = S_{II} + \bar{S}S_I$.

2.8 Parity of the Aberrations and the Principle of Symmetry

The aberrations can be divided into even and odd aberrations depending on the algebraic power of the aperture. The even aberrations are spherical aberration, astigmatism, and field curvature. The odd aberrations are coma and distortion. When there is some lens symmetry about the stop aperture, the odd aberrations tend to cancel, and this provides a mechanism to correct or mitigate the odd aberrations. This is known as the principle of symmetry about the stop.

Further Reading

Greivenkamp, J. *Field Guide to Geometrical Optics* (Bellingham, WA: SPIE Press, 2004).

Sasián, J. *Introduction to Aberrations in Optical Imaging Systems* (Cambridge, UK: Cambridge University Press, 2013).

3

Aspheric Surfaces

Optical systems comprise lenses and mirrors made with precise surfaces. Optical surfaces can be divided into spherical and nonspherical surfaces; the latter are called aspheric surfaces. For a given image quality, the choice of optical surfaces has a major impact on the packaging and cost of a lens system. Therefore, familiarity with types of optical surfaces, with how they can correct aberration, and with their manufacturing and testing methods is important in lens design. This chapter provides an overview of several useful surface types, some of their optical properties, and how they introduce and mitigate aberrations.

3.1 Spherical Surfaces

Spherical surfaces are the preferred optical surfaces because they are relatively easy to manufacture and are described by the equation, $r^2 = x^2 + y^2 + (z - r)^2$, where r is the radius of curvature. The reason for their ease of manufacturing is that two spherical surfaces of the same radius, one concave and one convex, fit each other regardless of their relative position. In traditional optics fabrication, two surfaces are rubbed against each other in the presence of an abrasive, and naturally they tend to conform to each other, acquiring a spherical form. Because of the ease of fabrication and testing, spherical surfaces are the default surfaces in lens design. Optically a spherical surface is specified by its radius of curvature, r, and its clear aperture.

Surfaces that are not spherical are called aspherical, and we assume that they have an axis of rotational symmetry. Specifically, their $sag(x^2 + y^2)$, or depth, z, depends on the radial distance, $\sqrt{x^2 + y^2}$, to the axis of rotation or the optical axis, which coincides with the z coordinate axis.

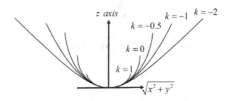

Figure 3.1 Cross-sections of conicoids with the same vertex radius of curvature and according to the conic constant k.

3.2 Conicoids

By rotating the conic sections about their axes, surfaces of revolution called conicoids, or conoids, are generated. These are described by their vertex radius of curvature, r, and their conic constant, $k = -\varepsilon^2$, where ε is the eccentricity. Unlike the sphere, the ellipsoid, the paraboloid, and the hyperboloid surfaces possess two separated optical foci. Light from a point source located at one focus, after reflection on the conicoid, converges to or appears to diverge from the other focal point.

In optical design, the sag of a conic surface is expressed by,

$$sag\left(x^2 + y^2\right) = z\left(x^2 + y^2\right) = \frac{c\left(x^2 + y^2\right)}{1 + \sqrt{1 - (1+k)(x^2 + y^2)c^2}}, \qquad (3.1)$$

where $c = 1/r$ is the vertex curvature of the surface. As shown in Figure 3.1, depending on the value of the conic constant k, the surface can be a sphere, $k = 0$; a paraboloid, $k = -1$; an ellipsoid, $-1 < k < 0$; a hyperboloid, $k < -1$; or a spheroid, $k > 0$. The equation of a conic is of second order, and it is possible to find the intersection point of a ray in closed mathematical form.

Within the fourth-order theory of aberrations, the correction of spherical aberration by a surface requires,

$$W_{040} = -\frac{1}{8}A^2 y\Delta\left(\frac{u}{n}\right) - \frac{1}{8}kc^3 y^4 \Delta(n) = 0. \qquad (3.2)$$

Then for an object at infinity we must have,

$$k = -\left(\frac{n}{n'}\right)^2. \qquad (3.3)$$

As shown in Figure 3.2, when $n' = 1$ there is no spherical aberration if $k = -n^2$, which requires a hyperboloid surface. When $n = 1$ there is no spherical aberration if $k = -1/n'^2$, which requires an ellipsoidal surface.

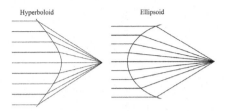

Figure 3.2 Left: Light refraction by a hyperboloid surface with $n = 1.5$, $n' = 1$, and $k = -2.25$. Right: Light refraction by an ellipsoid surface with $n = 1$, $n' = 1.5$, and $k = -0.4444$.

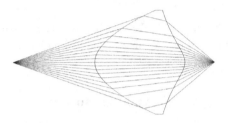

Figure 3.3 A lens with Cartesian surfaces forming a point image from a point object source.

For light reflection on a mirror surface to have no spherical aberration we must require,

$$k = -\left(\frac{1 + m}{1 - m}\right)^2, \qquad (3.4)$$

where m is the transverse magnification.

3.3 Cartesian Ovals

Cartesian ovals are curves or surfaces defined by using two points, one in object space, $s(0, 0, s)$, the other in image space, $s'(0, 0, s')$, and by requiring that the optical path length for any ray from the object to a surface point, $p(x_p, y_p, z_p)$, and to the image point be constant. Mathematically, Cartesian ovals are defined by,

$$n's' - ns = n'\sqrt{x_p^2 + y_p^2 + (z_p - s')^2} + n\sqrt{x_p^2 + y_p^2 + (z_p - s)^2}, \qquad (3.5)$$

where n and n' are the indices of refraction in object and image spaces, respectively. Cartesian ovals produce, geometrically, a perfect on-axis point image free from spherical aberration. Figure 3.3 shows a lens with Cartesian

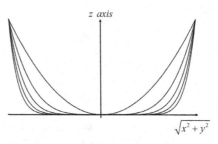

Figure 3.4 Monomials $x^2 + y^2$, $(x^2 + y^2)^2$, $(x^2 + y^2)^3$, $(x^2 + y^2)^4$, and $(x^2 + y^2)^5$.

ovals as surfaces. In some cases Cartesian ovals become conicoids, such as in reflection or when one of the defining points is at infinity.

3.4 Polynomial Surfaces

To extend the modeling capabilities, polynomial surfaces are also used in lens design. The sag of a polynomial surface is given by,

$$sag(x^2 + y^2) = A_2(x^2 + y^2) + A_4(x^2 + y^2)^2 + A_6(x^2 + y^2)^3 + \cdots, \quad (3.6)$$

where A_2, A_4, and A_6 are the second, fourth, and sixth-order coefficients of asphericity, respectively. Figure 3.4 shows the cross-sections of the monomials, $x^2 + y^2$, $(x^2 + y^2)^2$, $(x^2 + y^2)^3$, $(x^2 + y^2)^4$, and $(x^2 + y^2)^5$. Note that, as the algebraic order increases, most of the asphericity takes place toward the surface edge.

For many practical lens design problems, the superposition of a conicoid and a polynomial surface provides substantial flexibility to model optical surfaces, this is,

$$sag(x^2 + y^2) = \frac{c(x^2 + y^2)}{1 + \sqrt{1 - (1 + k)(x^2 + y^2)c^2}} + A_2(x^2 + y^2) + A_4(x^2 + y^2)^2 + A_6(x^2 + y^2)^3 + \cdots.$$

$$(3.7)$$

Usually the second-order coefficient of asphericity, A_2, is not used simultaneously with the vertex radius of curvature, $r = 1/c$, as first-order properties would depend on both r and A_2. The number of aspheric coefficients used depends on the polynomial convergence to find the ideal surface needed, and on the ability of the lens design optimizer to find a solution. In a polynomial surface the algebraic order of the monomials can be even, or even and odd.

Table 3.1 *Contributions to the Seidel sums from an aspheric surface with conic constant, k, and fourth-order coefficient of asphericity, A_4*

$\delta S_I = a$	$\delta S_{II} = \left(\frac{\bar{y}}{y}\right) a$	$\delta S_{III} = \left(\frac{\bar{y}}{y}\right)^2 a$
$\delta S_{IV} = 0$	$\delta S_V = \left(\frac{\bar{y}}{y}\right)^3 a$	$a = \left(-kc^3 + 8A_4\right) y^4 \Delta(n)$

Adding odd monomials to a surface description enhances its modeling capabilities; for example, a conical surface can be modeled with the term, $\sqrt{x^2 + y^2}$. The sag of an odd and even polynomial aspheric surface is,

$$sag\left(x^2 + y^2\right) = \frac{c(x^2 + y^2)}{1 + \sqrt{1 - (1 + k)(x^2 + y^2)c^2}} + A_1 \sqrt{x^2 + y^2} + A_2 \sqrt{x^2 + y^2}^2$$
$$+ A_3 \sqrt{x^2 + y^2}^3 + A_4 \sqrt{x^2 + y^2}^4 + A_5 \sqrt{x^2 + y^2}^5$$
$$+ A_6 \sqrt{x^2 + y^2}^6 + \cdots . \tag{3.8}$$

In the presence of a polynomial surface there is no closed form solution to the intersection point of a ray, and therefore an iterative procedure is used to find the intersection point to a high degree of accuracy. Aspheric surfaces can often lead to better optical performance, to size and weight reduction of an optical system, and, in some cases, to unique solutions to certain design problems. Depending on the application, aspheric surfaces can be cost effective, such as in plastic optical systems that are mass produced. For ease of fabrication and testing, spherical surfaces are the default surfaces to be specified, then conicoids, Cartesian ovals, and last polynomial surfaces. An aspheric surface is specified by its nominal vertex radius of curvature, conic constant, aspheric coefficients, and diameter.

3.5 Aberration Coefficients

The contributions to the Seidel sums from an aspheric surface specified with the vertex radius of curvature, $r = 1/c$, conic constant, k, and fourth-order coefficient of asphericity, A_4, are given in Table 3.1. An aspheric surface is thought of as the superposition of a sphere of radius r, and an aspheric cap defined by k and A_4. The fourth-order aberration contributed by the aspheric surface is the sum of the aberration by the spherical part, for example, S_I, for spherical aberration, and for the aspheric cap, δS_I.

The ratio of the chief ray height to the marginal ray height, $\bar{y}/y$, at the aspheric surface determines whether the surface will contribute only spherical aberration $(\bar{y}/y = 0)$, or in addition coma, astigmatism, and distortion $(\bar{y}/y \neq 0)$. When an aspheric surface is located at the aperture stop, or at a pupil, spherical aberration W_{040} is the only contribution of fourth-order.

3.6 Testing Aspheric Surfaces

Whenever an aspheric surface is specified it is necessary to determine how that surface could be tested. As an example, a paraboloid mirror has applications in astronomical telescopes, and a null corrector is often used for its testing. Light from a point source illuminates and passes through a lens system, called a null corrector, then it reaches the aspheric surface under test where the light is reflected, then passes a second time through the null corrector, and finally forms a point image. Any error on the surface of the mirror, called a figure error, produces an aberration in the point image and provides information about how to polish the mirror to correct its optical figure error. Since light passes twice through the null corrector, the configuration is referred to as a *double pass*.

An easy way to design a null corrector is in a *single pass*. As shown in Figure 3.5, light rays from a point at infinity are refracted by a paraboloid surface and become coincident with the normal lines to the paraboloid. This is modeled in a lens design program by setting the index of refraction prior to the paraboloid equal to $n = 1 \times 10^{-8}$. Since the index of refraction in object space is nearly zero, then the angle of refraction is nearly zero, and the refracted rays must coincide with the normal lines to the paraboloid. The refracted rays suffer from negative spherical aberration and form a ray caustic at the mirror vertex's center of curvature where a field lens is located. Then a relay lens is placed to form a point image. The negative spherical aberration from the paraboloid surface is compensated with the positive spherical aberration from the relay lens, which has spherical surfaces for ease of fabrication, characterization, and testing.

Near the ray caustic the ray height is non-linear, and for low $F/\#$ mirrors a single relay lens would not adequately compensate for aberrations, as also

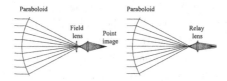

Figure 3.5 Testing a paraboloid mirror. Left: Schematic of an Offner null corrector. Right: Single relay lens null corrector.

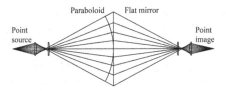

Figure 3.6 Schematic of an Offner null corrector in double pass. In modeling the test configuration with lens design software, a flat mirror has been added to unfold the path of light. This unfolding is not physically possible, but can be done in ray tracing.

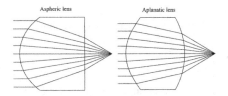

Figure 3.7 Left: An aspheric plano-convex lens corrected for spherical aberration. Right: A double convex aspheric lens corrected for spherical and coma aberration.

shown in Figure 3.5. The field lens redistributes the rays at the relay lens so that their height becomes more linear, allowing for a good match between the aberration from the paraboloid and the aberration from the relay lens. The field lens creates an image of the surface under test on the relay lens and controls higher order spherical aberration. Once a null corrector is designed in a single pass, it can be verified in a double pass, as shown in Figure 3.6.

3.7 Control of Spherical Aberration

Aspheric surfaces provide effective degrees of freedom to correct all orders of spherical aberration. Figure 3.7 shows a light focusing plano-convex lens free from spherical aberration. The convex surface uses an elliptical surface defined with a conic constant, k, and a fourth-order coefficient of asphericity, A_4. In this case only two coefficients are needed to provide satisfactory correction. However, if no conic constant is used, then up to fourteenth order aspheric coefficients are needed to properly correct for spherical aberration. This is due to the lack of fast convergence of the polynomial surface for this single lens design. For ease of alignment the lens must be corrected also for coma aberration so that a small field of view with excellent image quality is provided. This can be done by using the index of refraction as a second degree of freedom, or by using the curvature of the second surface, as also shown in Figure 3.7. When a lens system is corrected for both spherical aberration and coma aberration, it is referred to as an aplanatic lens.

Since spherical aberration depends on the fourth-order of the lens aperture, it is often best to correct for it at a location in a lens system where the marginal ray height is maximum. In addition, in order to not introduce other fourth-order aberrations, spherical aberration is often corrected by an aspheric surface that coincides, or is near, the aperture stop or a pupil.

3.8 Freeform Surfaces

The lens systems that are discussed in this book are mainly axially symmetric. However, if the axial symmetry in a lens system is not retained, new design possibilities and solutions to imaging problems result. Such non-axially symmetric lens systems also benefit from aspheric surfaces and, because these surfaces have at most one plane of symmetry and provide enhanced degrees of freedom to correct aberration, they are called freeform surfaces.

A freeform surface does not have axial or translational symmetry. Two cylindrical surfaces of the same radius fit to each other if they are translated along the cylinder axis, or translated perpendicular to the axis. Thus, cylindrical surfaces have translational symmetry, and have two orthogonal planes of symmetry. An axially symmetric surface has an infinite number of planes of symmetry, i.e., any meridional plane. Thus, a freeform surface is aspheric and has no more than one plane of symmetry.

Freeform surfaces are used in lens and mirror systems that do not have axial symmetry. These systems can have, in addition to spherical aberration, uniform astigmatism and uniform coma aberration over the field of view. The aberrations that plane symmetric systems can have are given in Appendix 4.

A useful freeform surface is defined by the superposition of a conic surface and a plane symmetric polynomial, symmetric about the Z–Y plane. The sag is,

$$sag\left(x^2, y\right) = \frac{c(x^2 + y^2)}{1 + \sqrt{1 - (1 + k)(x^2 + y^2)c^2}} + A_2\left(x^2\right) + A_3\left(x^2 + y^2\right)y$$
$$+ A_4\left(x^2 + y^2\right)^2 + \cdots. \tag{3.9}$$

The term, $A_2(x^2)$, adds a cylindrical deformation which is useful in controlling uniform astigmatism. The term, $A_3(x^2 + y^2)y$. is a cubic deformation which is useful for controlling uniform coma aberration. The term, $A_4(x^2 + y^2)^2$, is axially symmetric, and is useful for controlling spherical aberration.

3.9 User Defined Surfaces

For certain design problems, standard aspheric surfaces do not provide a solution because of a slow convergence and a limited number of aspheric terms. Then it is possible to write the code for a user defined surface to be used by lens design software. Writing the computer code for such a surface requires coming up with a potential surface type, producing the mathematical equations that describe that surface, finding the normal line at each surface point, performing refraction or reflection, and then compiling the code.

Further Reading

Brauneckeer, B., Hentschel, R., Tiziani, H. J. *Advanced Optics Using Aspherical Elements* (Bellingham, WA: SPIE Press, 2008).

Forbes, G. W. "Shape specification for axially symmetric optical surfaces," *Optics Express*, 15 (2007), 5218–26.

Greynolds, Alan W. "Superconic and subconic surface descriptions in optical design," Proceedings of SPIE 4832, International Optical Design Conference 2002 (December 23, 2002).

Hsueh, Chun-Che, Elazhary, Tamer, Nakano, Masatsugu, Sasián, José. "Closed-form sag solutions for Cartesian oval surfaces," *Journal of Optics*, 40(4) (2011), 168–75.

Offner, Abe. "A null corrector for testing paraboloidal mirrors," *Applied Optics*, 2(2) (1963), 153–55.

Reshidko, Dmitry, Sasián, José. "A method for the design of unsymmetrical optical systems using freeform surfaces," Proceedings of SPIE 10590, International Optical Design Conference 2017, 10590V (2017).

Sasián, José. "Design of null correctors for the testing of astronomical optics," *Optical Engineering*, 27(12) (1988), 121051.

Sasián, José, Reshidko, Dmitry, Li, Chia-Ling. "Aspheric/freeform optical surface description for controlling illumination from point-like light sources," *Optical Engineering*, 55(11) (2016), 115104.

Shultz, G. *Aspheric Surfaces*, Progress in Optics, Vol. XXV (Amsterdam: Elsevier, 1988), 349–415.

4

Thin Lenses

This chapter provides a discussion about thin lenses and how they are treated and analyzed for optical imaging. The concept of a thin lens is useful because aberration calculation with formulas is simplified. Structural aberration coefficients are used to determine aberrations and to show the rationale on the choice of the lens shape and aperture stop location. An understanding of how a singlet lens works is indispensable for the design of complex lens systems. The Wollaston periscopic lens, or landscape lens, is discussed regarding the technique of artificially flattening the field of view. A simple optical relay system is discussed, and then complexity is added to correct the primary monochromatic aberrations.

4.1 Thin Lens with the Aperture Stop at Lens

A useful concept in lens design is the thin lens where the central thickness is zero. As shown in Figure 4.1, from left to right, a thick plano-convex lens becomes a thin lens as the thickness becomes zero. The optical power, ϕ, of a thick lens in air is given by,

$$\phi = \phi_1 + \phi_2 - \frac{t}{n}\phi_1\phi_2,$$ (4.1)

where the surface optical powers are,

$$\phi_1 = \frac{n-1}{r_1}, \quad \phi_2 = -\frac{n-1}{r_2},$$ (4.2)

and where t is the central thickness, and n is the index of refraction of the lens material. By setting the thickness, t, equal to zero, we obtain a thin lens with optical power, $\phi = \phi_1 + \phi_2$, and, where the marginal and chief rays at each

Table 4.1 *Seidel sums in terms of structural aberration coefficients. Pupils located at principal planes*

$$S_I = \frac{1}{4} y_P^4 \phi^3 \sigma_I \quad S_{II} = \frac{1}{2} \mathcal{K} y_P^2 \phi^2 \sigma_{II} \quad S_{III} = \mathcal{K}^2 \phi \sigma_{III} \quad S_{IV} = \mathcal{K}^2 \phi \sigma_{IV} \quad S_V = \frac{2 \mathcal{K}^3 \sigma_V}{y_P^2}$$

Figure 4.1 From left to right, a thick lens becomes a thin lens with zero central thickness. A positive thin lens is represented in first-order optics as a double tipped arrowhead, as shown in the far right image.

surface are the same, this is $y_1 = y_2$ and $\bar{y}_1 = \bar{y}_2$. As a consequence, aberration calculation is simplified.

Table 4.1 gives the Seidel sums in terms of structural aberration coefficients, $\sigma_I, \sigma_{II}, \sigma_{III}, \sigma_{IV}, \sigma_V$, the Lagrange invariant, $\mathcal{K}$, the optical power, ϕ, of the lens or lens system, and the marginal ray height, y_P, at the principal planes. The structural aberration coefficients do not depend on the field of view, the optical speed, or the optical power, but on the structure of the lens system. Therefore, aberration properties are quantified as simply as possible. In addition, structural aberration coefficients allow lens system trade-off studies.

For a thin lens in air, the formulas for the structural coefficients, $\sigma_I, \sigma_{II}, \sigma_{III}, \sigma_{IV}, \sigma_V$, as a function of the shape, X, and conjugate, Y, factors, are given in Table 4.2. The shape factor, $X = (c_1 + c_2)/(c_1 - c_2)$, depends on the lens curvatures, and characterizes the shape of a lens. The conjugate factor, $Y = (1 + m)/(1 - m)$, depends on the transverse magnification at which the lens system works. Figure 4.2 illustrates the shape of a lens as a function of X. Changing the shape of a lens while maintaining its optical power is called lens bending. For equal lens curvatures the shape factor is not defined.

Spherical aberration, $W_{040} = S_I/8 = y_P^4 \phi^3 \sigma_I/32$, via the structural coefficient, σ_I, depends on the parameters, A, B, C, D, the shape factor, X, and the conjugate factor, Y, which in turn depends on the transverse magnification, m. Figure 4.3 (left) shows a plot of spherical aberration, W_{040}, in waves vs. the shape factor, X, for a thin lens with $f = 100$ mm, $F/4$, $Y = 1$, $\lambda = 587$ nm, and for various indices of refraction, and (right) difference between the actual *OPD* and the fourth-order spherical aberration, W_{040}. Note the strong dependence on the index of refraction, and the increased higher order aberration, OPD-W_{040}, for positive shape factors.

Table 4.2 *Structural aberration coefficients of a thin lens in air (Stop at lens)*

$$\sigma_I = AX^2 - BXY + CY^2 + D \qquad \sigma_{II} = EX - FY \qquad \sigma_{III} = 1 \qquad \sigma_{IV} = \frac{1}{n} \qquad \sigma_V = 0$$

$$A = \frac{n+2}{n(n-1)^2} \qquad B = \frac{4(n+1)}{n(n-1)} \qquad C = \frac{3n+2}{n} \qquad D = \frac{n^2}{(n-1)^2} \qquad E = \frac{n+1}{n(n-1)}$$

$$F = \frac{2n+1}{n} \qquad X = \frac{c_1 + c_2}{c_1 - c_2} = -\frac{r_1 + r_2}{r_1 - r_2} \qquad Y = \frac{1+m}{1-m} \qquad \phi = (n-1)\cdot\left(\frac{1}{r_1} - \frac{1}{r_2}\right)$$

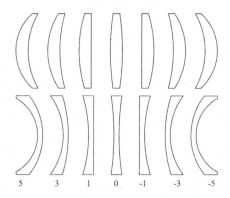

Figure 4.2 Shape of a lens from 5 to −5 for positive power lenses, upper row; and for negative power lenses, bottom row.

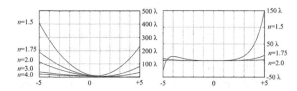

Figure 4.3 Left: thin lens spherical aberration, W_{040}, in waves at $\lambda = 587\,\text{nm}$ vs. shape factor, X, varying from −5 to 5, and for several indices of refraction. Right: higher-order, OPD-W_{040}, spherical aberration.

As a function of the shape factor, X, spherical aberration, σ_I, is quadratic. The plot of W_{040} vs. X is a parabola and, therefore, there may be two lens shapes for a given amount of spherical aberration. Spherical aberration strongly depends on the index of refraction; a higher index of refraction reduces both the amount of spherical aberration, W_{040}, and the amount of higher order, $OPD\text{-}W_{040}$, spherical aberration.

For a thin lens, the minimum spherical aberration, W_{040}, takes place when the shape factor is,

$$X = \frac{B}{2A} Y = \frac{2(n^2 - 1)}{2 + n} Y, \tag{4.3}$$

for which, σ_I becomes,

$$\sigma_I = \frac{n^2}{(n-1)^2} - \frac{n}{2+n} Y^2. \tag{4.4}$$

When the object is at infinity, $m = 0$, $Y = 1$, and the index is $n = 1.5$, the shape factor for minimum spherical aberration is $X \cong 0.71$ and $\sigma_I \cong 8.57$. Spherical aberration is positive for a thin lens with positive optical power, and negative when the optical power is negative.

Coma aberration, $W_{131} = S_{II}/2 = \mathcal{K} y_P^2 \phi^2 \sigma_{II}/4$, by way of σ_{II}, depends on the parameters, E, F, the shape factor, X, and the conjugate factor, Y. As a function of the shape factor, coma is a linear function and, therefore, there is a lens shape for which coma is zero,

$$X = \frac{F}{E} Y = \frac{(2n + 1)(n - 1)}{n + 1} Y. \tag{4.5}$$

When $n = 1.5$ and the object is at infinity, $Y = 1$, coma is $\sigma_{II} = (10X - 8)/3$. For a shape factor of $X = 0.8$, there is no coma, $\sigma_{II} = 0$. For $X = 1$, we have $\sigma_{II} = 2/3$. For a thin lens, spherical aberration and coma become simultaneously zero when $X = \pm(2n + 1)$ and $Y = \pm(n + 1)/(n - 1)$. For an index, $n = 1.5$, we have $X = \pm 4$ and $Y = \pm 5$. In this case, one surface satisfies $\Delta(u/n) = 0$, and the other surface satisfies $\bar{A} = 0$.

When the stop is at the thin lens, astigmatism aberration, $W_{222} = S_{III}/2 = \mathcal{K}^2 \phi/2$, does not depend on the shape or conjugate factor.

For a thin lens, Petzval field curvature, $W_{220P} = S_{IV}/4 = \mathcal{K}^2 \phi/4n$, does not depend on the shape or conjugate factors; it is inversely proportional to the lens index of refraction.

Distortion aberration is zero when the stop coincides with the thin lens, because the nodal points coincide with the lens vertex, and chief rays are not deviated.

Table 4.3 provide the structural aberration coefficients, σ_I and σ_{II}, for some special cases of a thin lens.

Table 4.3 *Thin lens structural aberration coefficients, σ_I and σ_{II}, for some special cases of the shape factor, X, and the conjugate factor, Y. Stop at lens*

X	Y	σ_I	σ_{II}
$X = \dfrac{2(n+1)(n-1)}{2+n} Y$	Y	$\sigma_I = \dfrac{n^2}{(n-1)^2} - \dfrac{n}{2+n} Y^2$ (minimum spherical aberration)	$\sigma_{II} = -\dfrac{1}{2+n} Y$
$X = \dfrac{(2n+1)(n-1)}{1+n} Y$	Y	$\sigma_I = \dfrac{n^2}{(n-1)^2} - \dfrac{n^2}{(n+1)^2} Y^2$	$\sigma_{II} = 0$ (zero coma)
$X = \pm(2n+1)$	$Y = \pm\dfrac{n+1}{n-1}$	$\sigma_I = 0$ (zero spherical aberration)	$\sigma_{II} = 0$ (zero coma)
$X = 0$ (equi-cx/cc lens)	Y	$\sigma_I = \dfrac{3n+2}{n} Y^2 + \dfrac{n^2}{(n-1)^2}$	$\sigma_{II} = -\dfrac{2n+1}{n} Y$
$X = 1$ (plano-cx/cc lens)	$Y = 1$ (Object at ∞)	$\sigma_I = 4\left(1 + \dfrac{2-n}{n(n-1)^2}\right)$	$\sigma_{II} = -2 + \dfrac{2}{n(n-1)}$
$X = -1$ (plano-cx/cc lens)	$Y = 1$ (Object at ∞)	$\sigma_I = \left(\dfrac{2n}{n-1}\right)^2$	$\sigma_{II} = -\dfrac{2n}{n-1}$

4.2 Thin Lens with Remote Aperture Stop

In the presence of aberration, aberrations change as the stop aperture shifts along the optical axis. The change of aberration due to stop shifting is given in Table 4.4 as a function of the stop shifting parameter, $\bar{S}_\sigma$.

In the presence of spherical aberration, the stop location can be used to correct coma aberration and, therefore, we must satisfy,

$$\sigma_{II}^* = 0 = \sigma_{II} + \bar{S}_\sigma \sigma_I, \tag{4.6}$$

which gives a stop shifting factor, $\bar{S}_\sigma$,

$$\bar{S}_\sigma = -\frac{\sigma_{II}}{\sigma_I}. \tag{4.7}$$

To simultaneously have zero astigmatism, we must also satisfy,

$$\sigma_{III}^* = 0 = \sigma_{III} + 2\bar{S}_\sigma \sigma_{II} + \bar{S}_\sigma^2 \sigma_I, \tag{4.8}$$

which requires,

$$\sigma_{II} = \pm\sqrt{\sigma_{III}\sigma_I} = \pm\frac{2n}{n-1}. \tag{4.9}$$

Therefore, the solution for zero coma and astigmatism aberration is a plano convex lens, $X = -1$, working at $Y = 1$, given that its coma aberration is,

$$\sigma_{II} = -\frac{2n}{n-1}. \tag{4.10}$$

The stop shifting parameter is,

$$\bar{S}_\sigma = \frac{y_P \bar{y}_P \phi}{2\mathcal{K}} = -\frac{\sigma_{II}}{\sigma_I} = \frac{n-1}{2n} = \frac{\bar{y}_P \phi}{2\bar{u}} = \frac{\bar{y}_P(n-1)}{2\bar{u}r_2}. \tag{4.11}$$

Then the distance, $\bar{s}$, to the stop from the thin lens is,

$$\bar{s} = \frac{\bar{y}_P}{\bar{u}} = \frac{r_2}{n}. \tag{4.12}$$

Table 4.4 *Stop-shifting formulas for structural coefficients*

$\sigma_I^* = \sigma_I$	$\sigma_{II}^* = \sigma_{II} + \bar{S}_\sigma \sigma_I$
$\sigma_{III}^* = \sigma_{III} + 2\bar{S}_\sigma \sigma_{II} + \bar{S}_\sigma^2 \sigma_I$	$\sigma_{IV}^* = \sigma_{IV}$
$\sigma_V^* = \sigma_V + \bar{S}_\sigma(\sigma_{IV} + 3\sigma_{III}) + 3\bar{S}_\sigma^2 \sigma_{II} + \bar{S}_\sigma^3 \sigma_I$	$\bar{S}_\sigma = \dfrac{y_P \bar{y}_P \phi}{2\mathcal{K}}$

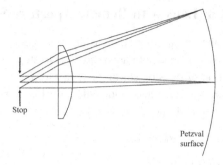

Figure 4.4 Plano-convex lens with stop in front and imaging on the Petzval surface.

Alternatively, we can understand this thin lens solution by considering that, for an object at infinity, we have $\Delta(u/n) = 0$ for the plane surface, and $\bar{A} = 0$ for the curved surface, which ensure both lens surfaces do not contribute any coma or astigmatism aberration. Figure 4.4 shows a thick plano-convex lens satisfying these conditions. In this simple lens, spherical aberration is controlled by reducing the aperture size, which has a strong impact because spherical aberration depends on the fourth power of the marginal ray height.

Field curvature aberration is given by,

$$W_{220} = \frac{1}{4}(S_{III} + S_{IV}).$$ (4.13)

Since there is no astigmatism, we have,

$$W_{220} = \frac{1}{4}S_{IV} = \frac{1}{4}\mathcal{H}^2 \sum_{i=1}^{j} \frac{n_{i+1} - n_i}{n_{i+1}n_i} \frac{1}{r_i}.$$ (4.14)

Then the image falls on the Petzval surface with vertex radius, ρ, given by the Petzval sum,

$$\frac{1}{\rho} = -\sum_{i=1}^{j} \frac{n_{i+1} - n_i}{n_{i+1}n_i} \frac{1}{r_i}.$$ (4.15)

For a system of thin lenses of power, ϕ_i, in air the Petzval sum becomes,

$$\frac{1}{\rho} = -\sum_{i=1}^{j} \frac{\phi_i}{n_i}.$$ (4.16)

For the singlet plano-convex lens the Petzval radius is $\rho = -nf'$.

4.3 Field Curves

The ideal image surface is a planar, flat surface. However, in the absence of astigmatism, the surface of sharp imaging is the Petzval surface. Therefore, it is desirable in a lens system to reduce the Petzval sum. If astigmatism aberration is present, the Petzval surface loses meaning. Then, for a small aperture, meridional rays focus on the tangential field curve, and sagittal rays focus on the sagittal field curve. A curve (not shown) between the sagittal and the tangential field curves is called the medial field curve. Figure 4.5 illustrates the field curves, and Table 4.5 gives the vertex curvature of the curves. The ideal image plane is labeled the Gaussian surface.

4.4 Optical Relay System

Creating an image of an object located at a finite distance or relaying an image to a given location is often required in optical systems. Thus, developing expertise about the design of optical relays is useful in optical engineering. A simple relay operating at $m = -1$ can be formed by combining two plano-convex lenses, as shown in Figure 4.6. The advantages are the simplicity, the correction for coma and astigmatism aberration, the control of spherical aberration by reducing the optical speed, and the relatively low cost.

Table 4.5 *Field curve vertex curvature in terms of structural coefficients*

$$C_{Petzval} = -n'\phi \cdot \sigma_{IV}$$

$$C_{Sagittal} = -n'\phi \cdot (\sigma_{IV} + \sigma_{III})$$

$$C_{Medial} = -n'\phi \cdot (\sigma_{IV} + 2\sigma_{III})$$

$$C_{Tangential} = -n'\phi \cdot (\sigma_{IV} + 3\sigma_{III})$$

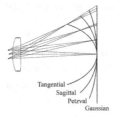

Figure 4.5 Field curves.

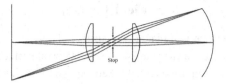

Figure 4.6 Optical relay consisting of two plano-convex lenses. The image lies on the Petzval surface.

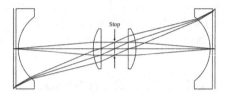

Figure 4.7 Symmetrical relay corrected for the monochromatic primary aberrations. An aspheric surface in a central glass plate coinciding with the aperture stop is used to correct for spherical aberration. The field lenses with negative focal length correct for Petzval field curvature.

If lens complexity is allowed, spherical aberration can be corrected with an aspheric surface located at the aperture stop. In addition, since spherical aberration, coma, and astigmatism depend on the marginal ray height at a thin lens, then no contribution to these aberrations would result from a lens placed at, or near, an image where the marginal ray height is zero. Such a lens at, or near, an image is known as a field lens, and would contribute Petzval field curvature aberration. Thus, by adding field flattener lenses, in addition to an aspheric plate at the stop, a relay corrected for the monochromatic primary aberrations would result, as shown in Figure 4.7. Distortion aberration is corrected because the relay is symmetrical about the aperture stop, and symmetrical about the imaging conjugates, given that $m = -1$.

4.5 Wollaston Periscopic Lens

Spectacle lenses have been around since the thirteenth century and, before the nineteenth century, the lenses used in spectacles were plano-convex or plano-concave in shape, likely for ease of fabrication. Using a plano lens for correcting the eye's vision provides a sharp image at the center of the visual field. However, toward the periphery of the field the images lose clarity. Wollaston noted that a circle looks the same in every direction when it is seen from its center, and reasoned that the best shape for the lenses in spectacles

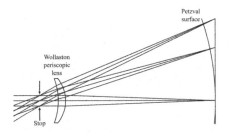

Figure 4.8 Meridional rays focus on the ideal image plane as negative astigmatism is introduced by the lens. In the absence of astigmatism, meridional and sagittal rays would focus on the Petzval surface.

was spherical. He had no optical theory to prove his assertion, but commissioned the fabrication of several lenses with the same focal length but of different optical shape. Wollaston then demonstrated that a meniscus lens shape was best to render a clear field of vision, and called these lenses periscopic, as they also allowed one to look clearly around the periphery of the field of view. In addition to being used in spectacles, periscopic lenses were used in camera obscuras to obtain brighter images (in comparison to camera obscuras without a lens) on a flat surface, as shown in Figure 4.8. The Wollaston periscopic lens is better known as the landscape meniscus lens.

The analysis of periscopic lenses made clear the importance of the aperture stop and its position in a lens system, and led to the concept of artificially flattening the field of view. To obtain sharp imaging on a flat surface there must be neither astigmatism aberration, W_{222}, nor Petzval field curvature aberration, W_{220P}. If there is no astigmatism, then the image falls on the Petzval surface. When the field is artificially flattened, negative astigmatism is introduced to flatten the tangential field curve.

Figure 4.9 shows the sagittal and tangential field curves when the ratio of astigmatism to Petzval field curvature is, $W_{222}/W_{220P} = -0.8$. For a given focal length, Wollaston periscopic lenses have two degrees of design freedom: the stop location and the lens shape. The lens shape is used to correct for coma aberration, and the stop position is used to introduce negative astigmatism, as given by stop shifting,

$$\sigma_{III}^* = \sigma_{III} + 2\bar{S}_\sigma \sigma_{II} + \bar{S}_\sigma^2 \sigma_I. \tag{4.17}$$

The lens contributes positive spherical aberration, σ_I, and the term $\bar{S}_\sigma^2 \sigma_I$ is positive. However, the lens contributes negative coma, σ_{II} (stop at lens), and the term $2\bar{S}_\sigma \sigma_{II}$ dominates, resulting in negative astigmatism to artificially flatten the field of view. Further, spherical aberration is controlled with the

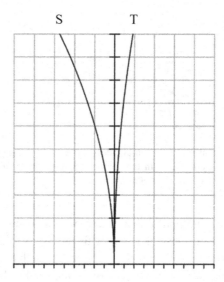

Figure 4.9 Astigmatic field curves for $W_{222}/W_{220P} = -0.8$. Meridional rays focus on the tangential field curve, and sagittal rays focus on the Sagittal field curve. The vertical axis represents the field of view, and the horizontal axis represents the optical axis.

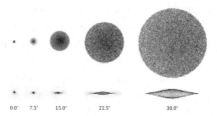

Figure 4.10 Top row, image quality as shown by spot diagrams for a plano-convex lens. Bottom row, image quality for a periscopic lens with a focal length of 100 mm, and a field of view from 0.0° to 30°. The spots are at the ideal image plane.

aperture stop diameter, and in such a simple lens there are no further degrees of freedom to correct chromatic aberrations, or distortion aberration. However, for a focal length of 100 mm, a field of view of ±30°, and an optical speed of *F*/16, a periscopic lens can render useful images for photography on a flat surface. Figure 4.10 (top row) shows spot diagrams for a plano-convex lens with the stop in front, and the bottom row shows spot diagrams for a periscopic lens with the stop in front. The spots are at the ideal image plane. Note that the periscopic lens provides a better imaging over the field of view, and the plano-convex lens provides a sharper image only at the field center.

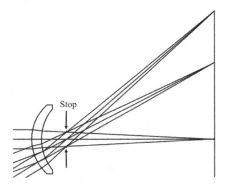

Figure 4.11 Periscopic lens with rear aperture stop. The lens index of refraction is $n = 2.42$, which reduces the Petzval field curvature.

There is another periscopic lens solution with the aperture stop in the rear, as shown in Figure 4.11. This solution has the advantage that the lens protects a possible shutter mechanism located at the aperture stop. However, the image quality is less favorable than when the aperture stop is in front. It is frequent in lens design to find several solutions with the same lens complexity.

4.6 Periskop Lens

The Wollaston periscopic lens suffers from distortion aberration. An improvement, although requiring adding lens complexity, is the Periskop lens in which two periscopic lenses are arranged symmetrically about the aperture stop. Because the odd aberrations, coma, and distortion, tend to cancel, they are substantially corrected in the Periskop lens, as shown in Figure 4.12.

Spherical aberration is mitigated by the aperture stop diameter to provide an $F/16$ optical speed. The field is artificially flattened to obtain a best image on a flat surface. Arranging lenses symmetrically, or nearly so, about the stop aperture is often used in lens design to help control odd aberrations. Periskop lenses have been used, for example, in overhead projectors.

4.7 Criterion for Artificially Flattening the Field

While the choice of negative astigmatism to artificially flatten the field of view was first done experimentally, we can now determine to fourth-order the ratio, W_{222}/W_{220P}, based on minimizing either the wavefront variance, σ_W^2, or the mean square spot size, $\bar{\varepsilon}^2$.

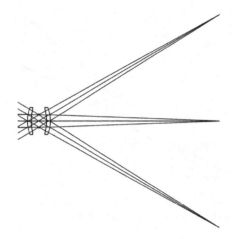

Figure 4.12 Periskop lens using the principle of symmetry to control odd aberrations.

Let us assume that the aberration function, $W\left(\vec{H},\vec{\rho}\right)$, is given by,

$$W\left(\vec{H},\vec{\rho}\right) = W_{020}\left(\vec{\rho}\cdot\vec{\rho}\right) + W_{040}\left(\vec{\rho}\cdot\vec{\rho}\right)^2 + W_{131}\left(\vec{H}\cdot\vec{\rho}\right)\left(\vec{\rho}\cdot\vec{\rho}\right)$$
$$+ W_{222}\left(\vec{H}\cdot\vec{\rho}\right)^2 + W_{220}\left(\vec{H}\cdot\vec{H}\right)\left(\vec{\rho}\cdot\vec{\rho}\right) + W_{311}\left(\vec{H}\cdot\vec{H}\right)\left(\vec{H}\cdot\vec{\rho}\right).$$
$$(4.18)$$

Then the variance of the wavefront is, $\sigma_W^2 = \overline{W^2} - \overline{W}^2$, where the mean square deformation is,

$$\overline{W^2} = \frac{1}{\pi}\int\limits_0^{2\pi}\int\limits_0^1 W^2\rho\,d\rho\,d\phi, \qquad (4.19)$$

and the mean deformation is,

$$\overline{W} = \frac{1}{\pi}\int\limits_0^{2\pi}\int\limits_0^1 W\rho\,d\rho\,d\phi. \qquad (4.20)$$

We can calculate the variance of the wavefront as,

$$\sigma_W^2 = \left(\begin{array}{l} \frac{1}{12}\left(W_{020} + W_{040} + \left(W_{220} + \frac{1}{2}W_{222}\right)\vec{H}\cdot\vec{H}\right)^2 + \frac{1}{180}W_{040}^2 \\ + \frac{1}{24}W_{222}^2\left(\vec{H}\cdot\vec{H}\right)^2 + \frac{1}{4}\left(\frac{2}{3}W_{131}\left|\vec{H}\right| + W_{311}\left(\vec{H}\cdot\vec{H}\right)\left|\vec{H}\right|\right)^2 + \frac{1}{72}W_{131}^2\,\vec{H}\cdot\vec{H} \end{array} \right).$$
$$(4.21)$$

By noting that $W_{220} = W_{220P} + W_{222}/2$, and by equating to zero the derivative of σ_W^2 with respect to W_{222}, we find that, to minimize the variance, we must have $W_{222} = -\frac{2}{3} W_{220P}$, which calls for a flat tangential field curve.

The mean square spot size, $\bar{\varepsilon}^2$, is given by,

$$\bar{\varepsilon}^2 = \frac{1}{n'^2 u'^2} \frac{1}{\pi} \int_0^{2\pi} \int_0^1 \left(\left(\frac{\partial W}{\partial \rho_x} \right)^2 + \left(\frac{\partial W}{\partial \rho_y} \right)^2 \right) \rho d\rho d\phi. \tag{4.22}$$

Then, for the aberration function, $W\left(\vec{H}, \vec{\rho} \right)$, we obtain,

$$\bar{\varepsilon}^2 = \frac{1}{n'^2 u'^2} \left(\begin{array}{c} 2 \left(W_{020} + \frac{4}{3} W_{040} + \left(W_{220} + \frac{1}{2} W_{222} \right) \vec{H} \cdot \vec{H} \right)^2 + \frac{4}{9} W_{040}^2 \\ + \frac{1}{2} W_{222}^2 \left(\vec{H} \cdot \vec{H} \right)^2 + \left(W_{131} \left| \vec{H} \right| + W_{311} \left(\vec{H} \cdot \vec{H} \right) \left| \vec{H} \right| \right)^2 + \frac{2}{3} W_{131}^2 \vec{H} \cdot \vec{H} \end{array} \right). \tag{4.23}$$

By equating to zero the derivative of $\bar{\varepsilon}^2$ with respect to W_{222}, we find that, to minimize the mean square spot size, we must have $W_{222} = -\frac{4}{5} W_{220P}$, which calls for a slightly backward curving tangential field, as shown in Figure 4.9.

Further Reading

Airy, G. "On the spherical aberration of the eyepieces of telescopes," *Cambridge Philosophical Transactions*, 2 (1827), 1–64.

Petzval, Joseph. *Bericht über die Ergebnisse einiger dioptrischen Untersuchungen* (Pesth: Verlag von Conrad Adolph Hartleben, 1943).

Sasián, J. *Introduction to Aberrations in Optical Imaging Systems* (Cambridge, UK: Cambridge University Press, 2013).

Wollaston, W. H. "On an improvement in the form of spectacle lenses," *Philosophical Magazine*, 17 (1804), 327–29.

Wollaston, W. H. "On a periscopic camera obscura and microscope," *Philosophical Transactions of the Royal Society of London*, 102 (1812), 370–77.

5

Ray Tracing

Ray tracing originated in optics to determine the path of light. However, ray tracing is used in modern technology by many fields, such as acoustics and computer graphics. Ray tracing is at the heart of optical design. Most optical calculations are done by tracing rays of light and, therefore, for competent lens design, it is important to have an understanding about how ray tracing is performed. This chapter provides an introduction to ray tracing, to ray tracing pitfalls, and to some useful ray tracing techniques.

5.1 Sequential Ray Tracing

We will consider an axially symmetric optical system. A plane passing through the axis of symmetry is called a meridional plane. Rays that are contained in the meridional plane are called meridional rays. Other rays that are not contained in a meridional plane, but that may intersect it, are called skew rays. Ray tracing of meridional rays is simple, as quantities are only required in two dimensions, while skew rays require quantities in three dimensions.

First-order rays are meridional rays and are traced assuming that the optical surfaces are flat but with optical power, ϕ. The first-order ray tracing equations $n'u' = nu - y\phi$ (refraction) and $y' = y + u't$ (transfer) are used to determine the intersection height of a ray with the next surface and the slope of the refracted ray.

Many calculations in a lens design program are done with first-order rays such as determining the ideal image position and size, the location of the pupils, and the cardinal points. First-order ray tracing requires a minimum computation time and is done very fast. Ray tracing time is critical, and minimizing it is important.

Real rays are traced applying Snell's law, $n' \sin(I') = n \sin(I)$, and determining the ray intersection with the actual specified surface to a high degree of accuracy. Most analyses in a lens design program depend on real ray tracing.

Real ray tracing is performed within a lens design program in an iterative process from surface to surface. For an imaging lens, the program is instructed to trace rays in a sequential manner so that a ray is traced first to surface #1, then to surface #2, and so forth till the last surface, which is often the image surface. Giving the ray starting coordinates, X_0, Y_0, and Z_0, in a previous surface, and the ray direction cosines, K, L, and M, of the ray at that point, the ray tracing algorithm determines the intersection point, X, Y, and Z, in the next surface, determines the normal line to the surface at that point, and performs reflection, refraction, or diffraction, to determine the new direction cosines, K', L', and M', of the ray. This process is repeated surface after surface in a sequential manner till the last surface is encountered. The optical path length of the ray is also computed.

The ray intersection with a conic surface can be determined in closed mathematical form, and this speeds up its ray tracing. Other aspheric surfaces such as polynomials require of an iterative process to determine the ray intersection point to a high degree of accuracy. Thus, ray tracing through aspheric surfaces takes more computing time.

Ray tracing data can be useful to check that a given lens design program is properly working, or that the designer properly interprets the results of the program.

5.2 Non-Sequential Ray Tracing

In some optical systems that contain many surfaces, it is not possible to know a priori, or to specify, the next surface that a ray will intersect. Then the ray tracing algorithm must determine which is the next surface to be intercepted. This requires more computing time, and significantly slows down the ray tracing. For example, the tracing of rays through a faceted gemstone, as shown in Figure 5.1, is performed with non-sequential ray tracing. Many non-imaging

Figure 5.1 Model of a faceted gemstone that retro reflects an incoming beam of parallel rays. Simulating light propagation in a gemstone requires non-sequential ray tracing.

and illumination optical systems are analyzed using non-sequential ray tracing. The design of lenses for imaging is usually done with sequential ray-tracing.

5.3 Ray Tracing Equations

Historically there have been many different sets of formulas for sequential ray tracing. For example, some formulas are suitable to be used with logarithm tables, some suitable for electronic computers, some for meridional rays or skew rays, some for spherical surfaces, and others for general aspheric surfaces.

To gain an idea of the computations needed for tracing a ray through a spherical surface, consider the coordinates of a skew ray, X_0, Y_0, and Z_0, and the direction cosines, K, L, and M, at a previous surface. If t is the axial distance to the next spherical surface, $X^2 + Y^2 + Z^2 - 2rZ = 0$, and r is its radius of curvature, then the coordinates of intersection of the ray, X, Y, and Z, with the next surface are given by, $X = KD + X_0$, $Y = LD + Y_0$, and $Z = MD + (Z_0 - t)$, where D is the length of the ray from the point, X_0, Y_0, and Z_0, to the point, X, Y, and Z. The distance, D, is given by,

$$D = \frac{-b \pm \sqrt{b^2 - 4ac}}{2a}, \tag{5.1}$$

where

$$a = 1, \tag{5.2}$$

$$b = -2r\left(M - \frac{KX_0 + LY_0 + MZ_0 - Mt}{r}\right), \tag{5.3}$$

and

$$c = r\left(\frac{(Z_0 - t)^2 + Y_0^2 + X_0^2}{r} - 2(Z_0 - t)\right). \tag{5.4}$$

The direction cosines of the normal line to the spherical surface at the point of intersection are, $k = -\frac{X}{r}$, $l = -\frac{Y}{r}$, and $m = 1 - \frac{Z}{r}$.

The cosines of the angle of incidence, I, and refraction, I', are,

$$\cos(I) = M - \frac{(KX + LY + MZ)}{r}, \tag{5.5}$$

and

$$\cos(I') = \sqrt{1 - \frac{n^2}{n'^2}(1 - \cos^2(I))}. \tag{5.6}$$

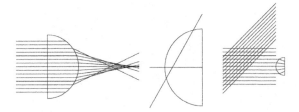

Figure 5.2 Ray tracing pitfalls. Left, total internal reflection for rays near the lens edge; center, two possible ray intersections at a convex surface; right, rays miss a convex spherical surface of small radius.

By defining the quantity, $G = \frac{n' \cos(I') - n \cos(I)}{r}$, the direction cosines, K', L', and M', of the refracted ray are found, $n'K' = nK - GX$, $n'L' = nL - GY$, and $n'M' = nM - G(Z - r)$. The optical path length for a number of j surfaces is given by, $OPL = \sum\limits_{i=1}^{j} n_i D_i$.

5.4 Ray Tracing Pitfalls

There are a number of ray tracing pitfalls to be aware of. These may interfere with proper system display and analyses. Ray total internal reflection may occur, and then the algorithm stops the ray tracing and may proceed to trace the next ray. There might be two possible intersection points with a given surface and directing the algorithm to the proper intersection point would be required. Missing a surface because the radius of curvature is too small is another occurrence. For some aspheric surfaces that are steep, the intersection point may not be found. Figure 5.2 shows (left) some rays suffering total internal reflection, (middle) a ray may intersect a surface at two points, and (right) rays may miss a surface.

5.5 Ray Definition

To trace a given ray requires a first starting point on the object surface, and a second point in the entrance or exit pupil plane of the system. By default, and for simplicity, the ray direction is computed based on first-order optics. Consequently, in object space, first-order rays and real rays perfectly coincide. However, at the exit pupil, or at the stop surface, first-order rays and real rays may not coincide. Then, through real ray aiming, the ray tracing algorithm may

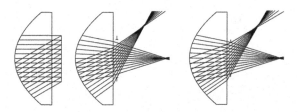

Figure 5.3 Ray aiming. Left, ray aiming to the entrance pupil; center, rays do not properly fill the stop aperture; right, rays properly fill the stop aperture because of ray aiming at the stop.

make both rays coincide at another surface rather than at the entrance pupil plane. Real ray aiming is necessary to properly simulate the passage of light rays through an optical system. Figure 5.3 shows (left) rays defined in object space by coordinates at the entrance pupil; (center) the traced rays through the lens do not coincide with the stop aperture because of pupil aberration; and (right) with ray aiming, rays are defined with coordinates at the stop and coincide with the stop aperture. However, ray aiming requires aiming each ray, and this can take significant time during lens optimization. Instead, ray definition can be achieved for proper filling of the aperture stop by setting light vignetting factors.

5.6 Reverse Ray Tracing

Nominally rays are traced from object space to image space. However, sometimes it is convenient to trace rays in reverse, from image space to object space. This can be done, for example, by reversing all the surfaces in the prescription of a lens. The initial design of a collimator lens, this is a lens that has the first-order image at infinity, can be carried by reverse ray tracing. Determining the presence of some ghost images and stray light in a lens system can be done by reverse ray tracing. Some calculations are enabled by reverse ray tracing and, thus, it is a useful technique in lens design.

Figure 5.4 illustrates how rays can enter a lens from high angles to be totally internally reflected by one surface and Fresnel reflected by the other surface of a lens to create a ghost image. Such ray paths can be discovered by reverse ray tracing. This is tracing rays from the image to the lens, performing reflection on one surface, then identifying total internal reflection on the other surface, and refracting the rays out of the lens to determine possible ray paths that contribute ghost images.

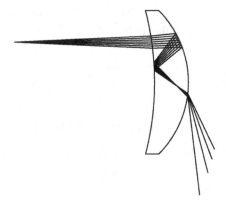

Figure 5.4 Reverse ray tracing to discover ray paths that can create ghost images. Rays are reverse ray traced from the image plane to a lens.

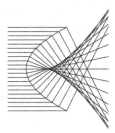

Figure 5.5 Parallel rays after refraction become normal to a parabola; the ray angle of refraction is zero. The index of refraction in object space is nearly zero.

5.7 Zero Index of Refraction

Lens design programs allow the index of refraction to be set to zero or nearly zero. This is not physically possible; however, by doing so, design simplification can be obtained in some lens designs, such as null correctors. If the index of refraction is set to zero in object space, while the index of refraction in image space is finite, then the angle of refraction must be zero, or nearly zero. Therefore, after refraction, a given ray will coincide with the normal line to the surface at the point of ray intersection. Figure 5.5 shows how rays after refraction become normal to a parabola.

5.8 Zero Dispersion

If the dispersion of a refracting material of a lens is set to zero (infinite v-number) then no chromatic aberration will be contributed by that lens. Then the contribution to chromatic aberration from other lenses in a system can easily be determined for analysis or further optimization. By setting a group or groups of lenses to have zero dispersion, the remaining groups can be corrected for chromatic aberration, while still accounting for monochromatic aberration from other lens groups. This is a method to selectively correct for chromatic aberration.

5.9 Infinite Index of Refraction

Ray diffraction can be simulated by setting the material index of refraction to infinity or, in practice, to a large value like $n = 10,000$. The optical power of a thin lens in air is given by,

$$\phi = (n - 1) \left(\frac{1}{r_1} - \frac{1}{r_2} \right). \tag{5.7}$$

As the index of refraction increases and the optical power remains the same, the difference in radii of curvature decreases, then the lens tends to behave as a diffractive optical element working at the first order, $m = 1$.

Figure 5.6 shows an aplanatic diffractive optical element on a spherical substrate modeled as a thin lens with an index of refraction, $n = 10,000$, with $r_1 = 100$, $r_2 = 100.01$, and $t = 0$. Rays from an object at infinity are focused by the aplanatic element.

5.10 Negative Thickness

The ray tracing algorithm allows rays to trace when the thickness to the next surface is negative. Figure 5.7 shows a double convex lens and several real rays traced. The marginal ray first intersects the front surface of the lens, in sequence. However, the distance to the edge of the rear surface is negative.

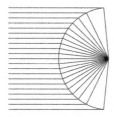

Figure 5.6 Aplanatic element modeled with a high index of refraction, $f' = 100$. Parallel rays diffract at a spherical in shape, diffractive optical element defined with two surfaces, and an index of refraction of $n = 10,000$.

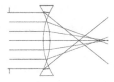

Figure 5.7 A marginal real ray traces backwards when the distance to the next surface is negative.

Figure 5.8 Rays are traced to the aperture stop, and then to the first surface of the lens system. A negative thickness is set for the aperture stop surface to position the first surface of the lens to the left of the aperture stop aperture.

Consequently, the marginal ray, after refracting in the front surface, intersects the rear surface by traveling backwards, and then refracts in the rear surface to proceed forward. The negative edge thickness is not physically possible, but real ray tracing is still performed.

5.11 Floating the Aperture Stop

Sometimes the best position for the aperture stop in a lens system needs to be identified. If the stop aperture is defined as the first surface of the lens system, and a negative thickness is set for this surface, then the internal position of the stop in a lens can be identified by allowing the lens design optimizer to vary the negative thickness. As shown in Figure 5.8, with this method light rays are traced to the aperture stop, which defines and coincides with the entrance pupil, then rays are reverse traced to the first surface of the lens system, and finally rays are traced forward through the lens. Once the best position of the entrance pupil is identified, the lens internal stop aperture location is set.

5.12 Dummy Surfaces

Dummy surfaces are surrounded by the same media (same index of refraction, before and after, for all wavelengths) and, thus, they do not cause ray refraction. They are used in a lens design layout to establish reference distances and to calculate beam characteristics at specific system locations.

5.13 Index Interpolation

For accurate ray tracing it is necessary to know the index of refraction at a given wavelength, λ, for the lens material being used. The glass manufacturer measures the index of refraction at a number of wavelengths, and makes a fit to an interpolation formula to predict the index of refraction at other wavelengths. There have been several formulas used in the past, but currently the Sellmeier formula is the most common, as its derivation is based on physical principles and can be accurate over a relatively wide bandwidth. This formula is:

$$n^2(\lambda) = 1 + \frac{B_1\lambda^2}{\lambda^2 - C_1} + \frac{B_2\lambda^2}{\lambda^2 - C_2} + \frac{B_3\lambda^2}{\lambda^2 - C_3}, \tag{5.8}$$

where B and C are Sellmeier coefficients.

An optical engineer must make sure that the optical design program properly determines the index of refraction of the material to be used in a lens; otherwise ray tracing can be inaccurate. Depending on the lens system application, the ambient temperature and atmospheric pressure may need to be considered in the lens design, since the index will change slightly as temperature and pressure change.

Another dispersion formula that is useful is the Schott formula:

$$n^2(\lambda) = A + A_1\lambda^2 + A_2\lambda^{-2} + A_4\lambda^{-4} + A_6\lambda^{-6} + \cdots. \tag{5.9}$$

By setting the Schott formula to $n(\lambda) = (10,000)\lambda$, a dispersive diffractive optical element can be modeled with standard ray tracing.

Further Reading

Allen, W. A., Snyder, J. "Ray tracing through uncentered and aspheric surfaces," *Journal of the Optical Society of America*, 42(4) (1952), 243–49.

Feder, D. P. "Optical calculations with automatic computing machinery," *Journal of the Optical Society of America*, 41(9) (1951), 630–35.

Ford, P. W. "New ray tracing scheme," *Journal of the Optical Society of America*, 50 (1960), 528–33.

Freniere, Edward R., Tourtellott, John. "Brief history of generalized ray tracing," Proceedings of SPIE 3130, Lens Design, Illumination, and Optomechanical Modeling (1997); doi: 10.1117/12.284059.

Liping, Z., Minxtan, W., Guofan, J., Weizhen, Y. "Ray tracing through arbitrary DOE based on Fermat's principle," *SPIE*, 3130 (1997), 238–44.

Sasián, J. M., Lerner, S. A., Lin, T. Y., Laughlin, L. "Ray and Van Citter-Zernike characterization of spatial coherence," *Applied Optics*, 40(7) (2001), 1037–43.

Spencer, G. H., Murty, M. V. R. K. "General ray-tracing procedure," *Journal of the Optical Society of America*, 52 (1962), 672–78.

Sweatt, W. C. "Describing holographic optical elements as lenses," *Journal of the Optical Society of America*, 67 (1977), 803–8.

Wang, Y., McDonald, J. "Ray tracing and wave aberration calculation for diffractive optical elements," *Optical Engineering*, 35(7) (1996), 2021–26.

6

Radiometry in a Lens System

Lens systems produce images by transferring radiant energy. At any plane transverse to the optical axis in an optical system there is a light distribution that may be subject to specifications. The light distribution is modeled with the laws of radiometry. To have a broader understanding about how optical systems work it is relevant to discuss how radiometric aspects impact the design of a lens system. In particular, we are concerned with the light distribution at the exit pupil and image planes. This chapter discusses basic and useful radiometric concepts in a lens system.

6.1 The Pinhole Camera

Let us consider a pinhole camera, as shown in Figure 6.1, where ds' is an element of area of the pinhole, and s' is the distance to the observation plane along the optical axis. At the observation plane there is an element of area, da'. Behind the pinhole aperture, which may be large in diameter, there is a Lambertian source of radiance, L_0', in units of W/m²-sr. The element of optical flux, $d\phi'^2$, in Watts, transferred to the element of area, da', is:

$$d\phi'^2 = L_0' \frac{ds' da'}{s'^2} \cos^4(\theta'). \tag{6.1}$$

Then the element of irradiance, dI', in W/m², at any point in the observation plane is given by,

$$dI' = L_0' \frac{ds'}{s'^2} \cos^4(\theta'), \tag{6.2}$$

where θ' is the angle with respect to the optical axis of the line defined by ds' and da'.

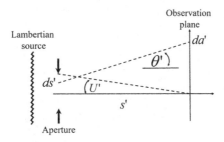

Figure 6.1 Geometry of a pinhole camera.

The irradiance in the optical axis, I_0', is obtained by summing over all points of the pinhole aperture, which is assumed to be circular and to have a radius, a. This is given by,

$$I_0' = 2\pi \frac{L_0'}{s'^2} \int_0^a r \left(\frac{r}{\sqrt{r^2 + s'^2}} \right)^4 dr = 2\pi L_0' \int_0^{U'} \sin(\xi) \cos(\xi) d\xi$$

$$= \pi \left(\frac{L_0'}{n'^2} \right) n'^2 \sin^2(U'). \tag{6.3}$$

The quantity $\frac{L_0'}{n'^2} = cte$ is known as the radiance theorem, and remains invariant in an optical system that does not lose light; for example, by material absorption. Therefore, the on-axis irradiance, I_0', is proportional to the square of the index of refraction of the observation media and proportional to the square of the sine of the semi-angle subtended by the pinhole aperture. The closer the observation point is to the pinhole camera, the larger the irradiance, I_0', becomes.

For radiometric purposes and to first-order, light from the exit pupil of a lens system can be modeled like that of a pinhole camera. Then, the on-axis irradiance at the image plane of a lens system is given by,

$$I_0' = \pi \frac{L_0'}{n'^2} NA^2, \tag{6.4}$$

where $NA = n' \sin(U')$ is the numerical aperture of the lens system in image space.

Using the approximation,

$$\sin^2(U') \simeq \frac{1}{4(F/\#)^2}, \tag{6.5}$$

in the relationship for I_0', allows one to conclude that the image brightness in the optical axis is inversely proportional to the square of the $F/\#$ at which an

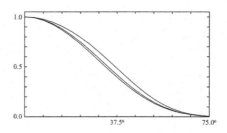

Figure 6.2 Relative illumination of pinhole cameras at $F/100$ (bottom curve), $F/2$ (intermediate curve), and at $F/1$ (top curve). The field of view is in degrees in the horizontal axis from $0°$ to $75°$.

objective lens is working. Thus, to cast bright images, lenses with small $F/\#$s are required.

6.2 Pinhole Camera Relative Illumination

If the aperture is small, i.e., a pinhole, and has an area ds', then, for an off-axis observation point, the element of irradiance is given by,

$$dI' = \frac{L_0' ds'}{s'^2} \cos^4(\theta').$$
(6.6)

The relative illumination at the observation plane is defined as the ratio of the off-axis irradiance to the on-axis irradiance, and is simply given by,

$$RI = \cos^4(\theta').$$
(6.7)

This relation is known as the cosine^4 law of illumination, and serves as a reference to describe the relative illumination of a lens system.

A pinhole can have different sizes and, in consideration of the distance to the observation plane, different $F/\#$'s can be defined. Figure 6.2 shows the relative illumination of pinhole cameras operating at $F/100$, $F/2$, and $F/1$. As can be seen, there is little difference in the relative illumination when above $F/2$.

The transverse position of the pinhole with respect to the optical axis is specified by $\vec{\rho}$, and the observation point is specified by $\vec{H}$. Then the relative illumination is given in terms of the first-order marginal, u', and chief, $\bar{u}'$, ray slopes as,

$$RI\left(\vec{H}, \vec{\rho}\right) = \cos^4(\theta')$$
$$= 1 - 2u'^2\left(\vec{\rho} \cdot \vec{\rho}\right) - 4u'\bar{u}'\left(\vec{H} \cdot \vec{\rho}\right) - 2\bar{u}'^2\left(\vec{H} \cdot \vec{H}\right) + \cdots$$
(6.8)

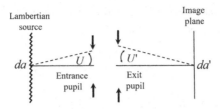

Figure 6.3 Representation of an optical system imaging a Lambertian source.

This relation serves as a reference to describe irradiance changes in an optical system.

6.3 Ratio of On-Axis Irradiance to Exitance in an Optical System

Let us consider now a lens system imaging a Lambertian source of radiance, L_0. The aperture stop and the entrance pupil coincide in object space. If da is an element of area on the optical axis, then the element of optical flux emitted by the source is given by,

$$d\phi = 2\pi L_0 da \int_0^U \sin(\xi) \cos(\xi) d\xi = \pi L_0 da \sin^2(U),\qquad(6.9)$$

and an on-axis exitance in object space is given by,

$$I_0 = \pi L_0 \sin^2(U).\qquad(6.10)$$

The angle, U, is the semi-angle subtended by the entrance pupil, as shown in Figure 6.3.

The ratio of the on-axis irradiance to exitance in the object and image planes is then,

$$\frac{I_0'}{I_0} = \frac{L_0'}{L_0} \frac{\sin^2(U')}{\sin^2(U)}.\qquad(6.11)$$

Recalling the radiance theorem, which states that the ratio of radiances in different media is equal to the ratio of the square of the indices of refraction,

$$\frac{L_0'}{L_0} = \frac{n'^2}{n^2},\qquad(6.12)$$

and expressing the real ray angles, U and U', as a function of the marginal and chief first-order ray slopes, allows us to write,

$$\frac{I_0'}{I_0} = \frac{n'^2}{n^2}\frac{\sin^2(U')}{\sin^2(U)} \simeq \frac{n'^2}{n^2}\frac{1+u^2}{u^2}\frac{(u'+\Delta u')^2}{1+(u'+\Delta u')^2} \simeq \frac{1}{m^2}\left(1+2\frac{W_{131}}{Ж}\right), \quad (6.13)$$

where the transverse magnification is m, and where we have used the relationships $\frac{\Delta u'}{u'} = \frac{\overline{W}_{311}}{Ж}$ and $\overline{W}_{311} = W_{131} + \frac{1}{2}Ж\Delta(u^2)$. The increment in slope, $\Delta u'$, is included as the marginal real ray may not coincide with the first-order marginal ray in image space.

If the optical system obeys the sine condition, $u\sin(U') = u'\sin(U)$, there is no coma, $W_{131} = 0$, and the ratio of the on-axis irradiance to exitance is inversely proportional to the square of the transverse magnification. In the presence of coma aberration, W_{131}, or equivalently in the presence of pupil distortion, $\overline{W}_{311}$, the ratio of on-axis irradiance to exitance changes. Effectively, the exit pupil increases or decreases in size, changing U', and then making the on-axis image point brighter or dimmer.

6.4 Lens System Relative Illumination

Let us consider an optical system where the aperture stop and the entrance pupil coincide. The exit pupil, being the image of the aperture stop in image space, can be distorted, not only for the on-axis beam but for off-axis beams. As the exit pupil distorts the image irradiance changes and, therefore, the relative illumination does too. To a second-order of approximation, the relative illumination, RI, of a system can be written as,

$$RI\left(\vec{H}\right) = 1 - \left(2\bar{u}'^2 - \frac{4}{Ж}\overline{W}_{131}\right)\left(\vec{H}\cdot\vec{H}\right). \quad (6.14)$$

In the absence of pupil coma, $\overline{W}_{131}$, the relative illumination is that of a pinhole camera. In the presence of pupil coma, the exit pupil changes size and is anamorphically distorted. This distortion changes the convergence of the focusing beams and, thereby, the irradiance. Because pupil coma is related to image distortion, $\overline{W}_{131} = W_{311} + \frac{1}{2}Ж\Delta(\bar{u}^2)$, the change in relative illumination can also be explained as a result of the decrease or increase of the imaging area for a given amount of optical flux.

If the stop is located at the exit pupil, then the relative illumination is, to second-order, that of a pinhole camera,

$$RI\left(\vec{H}\right) = 1 - 2\bar{u}'^2\left(\vec{H}\cdot\vec{H}\right). \quad (6.15)$$

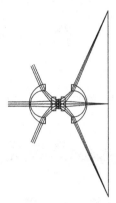

Figure 6.4 A wide angle objective lens following a cosine^3 illumination law.

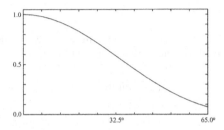

Figure 6.5 Relative illumination of the Roossinov lens. The field of view is in degrees in the horizontal axis from 0° to 65°.

Because pupil coma can affect the relative illumination, some wide-angle lenses take advantage to compensate for the cosine^4 law of illumination. The Roossinov lens of 1950 (US Patent 2,516,724), as shown in Figure 6.4, was designed as a nearly symmetrical lens where each half contributes a large amount of pupil coma which nearly cancels by symmetry. However, because the aperture stop is in the middle of the lens, only pupil coma that is contributed by the second half of the lens counts for the lens relative illumination; this pupil coma makes the lens to have a cosine^3 law of illumination, as shown in Figure 6.5.

The exit pupil as a function of the field of view is increasingly magnified and anamorphically distorted, as shown in Figure 6.6.

6.5 Light Vignetting

The aperture stop defines the relative aperture of the on-axis beam. It also defines the aperture for off-axis beams. However, there might be other

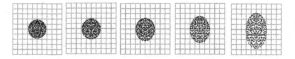

Figure 6.6 Spot diagrams showing the cross-section of different off-axis beams at the exit pupil plane. From left to right, on-axis to off-axis field positions.

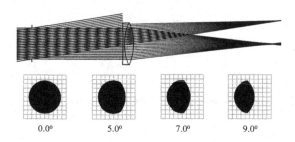

$\quad$ 0.0°$\qquad$ 5.0°$\qquad$ 7.0°$\qquad$ 9.0°

Figure 6.7 For off-axis beams the physical size of the lens aperture limits the amount of light. The loss of light increases as the field of view increases.

apertures in an optical system that limit the amount of light for off-axis beams. Rays of light can be obstructed, or clipped, for example by the physical size of a lens. This loss of light in an off-axis beam is known as light vignetting. Figure 6.7 shows how beam clipping can take place for off-axis beams. Because there is light loss, the relative illumination decreases.

However, in some lenses, light vignetting is introduced on purpose to suppress rays that have aberration beyond tolerances. Light vignetting has been used extensively in photographic lenses to reduce aberration at low $F/\#$s and to reduce the overall lens diameter and weight.

6.6 Irradiance at the Exit Pupil Plane

It is of interest to describe the irradiance at the exit pupil plane of a lens system. We assume that the stop aperture coincides with the exit pupil, and that the object is a Lambertian source that is circular to maintain axial symmetry. The function,

$$\bar{I}\left(\vec{H}, \vec{\rho}\right) = \bar{I}_{000} + \bar{I}_{200}\left(\vec{\rho} \cdot \vec{\rho}\right) + \bar{I}_{111}\left(\vec{H} \cdot \vec{\rho}\right) + \bar{I}_{020}\left(\vec{H} \cdot \vec{H}\right) \ldots, \quad (6.16)$$

provides the irradiance at the exit pupil as a function of the field, $\vec{H}$, and aperture, $\vec{\rho}$, vectors. Figure 6.8 shows graphically the second-order terms $\bar{I}_{200}\left(\vec{\rho} \cdot \vec{\rho}\right)$, $\bar{I}_{111}\left(\vec{H} \cdot \vec{\rho}\right)$, and $\bar{I}_{020}\left(\vec{H} \cdot \vec{H}\right)$, which represent irradiance aberrations.

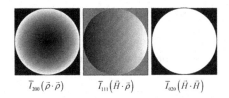

$$\bar{I}_{200}\left(\vec{\rho}\cdot\vec{\rho}\right) \qquad \bar{I}_{111}\left(\vec{H}\cdot\vec{\rho}\right) \qquad \bar{I}_{020}\left(\vec{H}\cdot\vec{H}\right)$$

Figure 6.8 Second-order terms of irradiance changes at the exit pupil of a lens system.

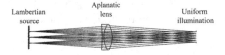

Figure 6.9 An aplanatic lens can be used to produce uniform illumination from a Lambertian source.

With $\bar{I}_{000} = 1$, the second-order coefficients in terms of image aberration coefficients are,

$$\bar{I}_{020}\left(\vec{H}\cdot\vec{H}\right) = \left[-2\bar{u}'^{2} - \frac{4}{\mathcal{K}}W_{311}\right]\left(\vec{H}\cdot\vec{H}\right), \qquad (6.17)$$

$$\bar{I}_{111}\left(\vec{H}\cdot\vec{\rho}\right) = \left[-4u'\bar{u}' - \frac{4}{\mathcal{K}}W_{220} - \frac{6}{\mathcal{K}}W_{222}\right]\left(\vec{H}\cdot\vec{\rho}\right), \qquad (6.18)$$

and

$$\bar{I}_{200}\left(\vec{\rho}\cdot\vec{\rho}\right) = \left[-2u'^{2} - \frac{4}{\mathcal{K}}W_{131}\right]\left(\vec{\rho}\cdot\vec{\rho}\right). \qquad (6.19)$$

If a lens system is free from coma aberration, $W_{131} = 0$, and collimates light, $u' = 0$, then the illumination at the exit pupil would be uniform. This comes about because, for an axially symmetric object, the contributions of the term $\bar{I}_{111}\left(\vec{H}\cdot\vec{\rho}\right)$ will cancel, i.e., $\bar{I}_{111}\left(\vec{H}\cdot\vec{\rho}\right) + \bar{I}_{111}\left(-\vec{H}\cdot\vec{\rho}\right) = 0$, the term $\bar{I}_{020}\left(\vec{H}\cdot\vec{H}\right)$ does not depend on the aperture vector, $\vec{\rho}$, at the exit pupil, and the term $\bar{I}_{200}\left(\vec{\rho}\cdot\vec{\rho}\right)$ will vanish.

An aplanatic lens, as shown in Figure 6.9, will produce uniform illumination at the exit pupil when the object is a Lambertian source. If the lens is aplanatic, but is imaging $u' \neq 0$, like a microscope objective used in reverse, the irradiance at the exit pupil would follow a cosine[4] law to second-order.

6.7 Optical Étendue

The optical flux, ϕ, transferred from a source of radiance, L_0 (here assumed uniform), and area, A_s, to an optical component or lens system is given by,

$$\phi = \frac{L_0}{n^2}\varepsilon, \qquad (6.20)$$

where ε is the optical étendue associated with the optical component or system. Two first-order approximations to the étendue are $\varepsilon = \pi n^2 A_s \sin^2(U)$ and $\varepsilon = \pi^2 \mathcal{K}^2$. The étendue gives further meaning to the Lagrange invariant, $\mathcal{K}$, and it is a measure of the capacity of an optical component or lens system to transfer optical flux.

Every optical element or system has associated an étendue and, in order to avoid loss of light, the étendue of all the components of a system must be equal or greater than the étendue that defines the light gathering properties of the system. For example, the pixel size of a sensor often defines the $F/\#$ of an imaging system, and, given the sensor size area, the minimum étendue of the system is established. Then all components in the system must have at least the sensor's étendue to avoid light loss and fully illuminate the sensor.

As another example, consider a sharp imaging fisheye lens with a semi-field of view, θ, and with an entrance pupil radius, y_e. The etendue, ε, of the system in object space is $\varepsilon = \pi^2 y_e^2 \sin^2(\theta)$. Assuming telecentricity in image space, $\bar{u}' = 0$, and the radius of the image to be $\bar{Y}_i$, then we have that, whenever $\sin(U') \cong \frac{1}{2F/\#}$, the étendue in image space is $\varepsilon' = \frac{\pi^2 \bar{Y}_i^2 y_e^2}{f^2}$, where f is the focal length of the lens. If there is no light loss, we must have $\varepsilon' = \varepsilon$, which leads to the lens mapping relationship, $\bar{Y}_i = f \sin(\theta)$. This mapping differs from the ideal, $\bar{Y}_i = f \tan(\theta)$, mapping that many lenses produce.

Still another example of the use of étendue is in deriving the sine condition. The optical flux emitted from a Lambertian source is given by, $\phi = L_0 \pi A_s \sin^2(U)$. The flux forming a sharp image of the source is given by, $\phi' = \phi = L_0' \pi A_s' \sin^2(U')$. Using the radiance theorem, we obtain, $n^2 A_s \sin^2(U) = n'^2 A_s' \sin^2(U')$, which, after using, $n\bar{y}_o u = n'\bar{y}_i u'$, simplifies into the sine condition, $u' \sin(U) = u \sin(U')$. A discussion of the sine condition is given in Appendix 5.

Further Reading

Gardner, C. "Validity of the cosine-fourth-power law of illumination," *Journal of Research of the National Bureau of Standards*, 39 (1947), 213–19.

Kingslake, R. *Illumination in Optical Systems, Applied Optics and Optical Engineering*, Vol. II (New York: Academic Press, 1965).

Koshel, R. J. *Illumination Engineering: Design with Non-Imaging Optics* (Hoboken, NJ: Wiley-IEEE Press, 2013).

Palmer, J. M., Grant, B. G. *The Art of Radiometry* (Bellingham, WA: SPIE Press, 2010).

Reiss, M. "The $\cos^4$ law of illumination," *Journal of the Optical Society of America*, 35 (4) (1945), 283–88.

Reiss, M. "Notes on the $\cos^4$ law of illumination," *Journal of the Optical Society of America*, 38(11) (1948), 980–86.

Reshidko, Dmitry, Sasián, José. "Geometrical irradiance changes in a symmetric optical system," *Optical Engineering*, 56(1) (2017), 015104.

Rimmer, M. P. "Relative illumination calculations," *Proceedings of SPIE*, 655 (1986), 99–104.

Roossinov, M. M. Wide angle orthoscopic anastigmatic photographic objective, US Patent 2,516,724 (1950).

Sasián, J. *Introduction to Aberrations in Optical Imaging Systems* (Cambridge, UK: Cambridge University Press, 2013).

Siew, Ronian. "Distinction between image magnification and irradiance magnification: a commentary," *Optical Engineering*, 56(2) (2017), 029701.

7

Achromatic and Athermal Lenses

The index of refraction of glass depends on the wavelength of light. For N-BK7 glass from Schott Company, the index of refraction is shown in Figure 7.1 for wavelengths ranging from 0.4 to 0.8 µm; shorter wavelengths have a higher index of refraction than longer wavelengths. Since the angle of refraction depends on the index of refraction, then the angle of refraction varies as the wavelength changes. This results in chromatic aberration. Similarly, the index of refraction and the radii of curvature and thickness of a lens vary with changes in lens temperature. This results in thermal aberrations; thermal change of focus, and thermal change of magnification. This chapter discusses both types of aberrations and their correction.

7.1 Chromatic Change of Focus and Magnification

When the wavelength of light changes, the aberration function of a system also changes and, to second-order, it can be written as,

$$W\left(\vec{H},\vec{\rho}\right) = \partial_\lambda W_{000} + \partial_\lambda W_{020}\left(\vec{\rho}\cdot\vec{\rho}\right) + \partial_\lambda W_{111}\left(\vec{H}\cdot\vec{\rho}\right) + \partial_\lambda W_{200}\left(\vec{H}\cdot\vec{H}\right),$$

$$(7.1)$$

where $\partial_\lambda W_{020}$ is the coefficient for the chromatic change of focus and $\partial_\lambda W_{111}$ is the coefficient for the chromatic change of magnification. These aberrations are known in the lens design literature as axial/longitudinal and lateral/transverse chromatic aberrations, respectively. The chromatic piston terms, $\partial_\lambda W_{000}$ and $\partial_\lambda W_{200}$, are neglected as they do not degrade the image quality. As shown in Figure 7.2 (left), in the presence of chromatic change of focus, a marginal ray is dispersed and intersects the optical axis at different distances along the optical axis, and (right), in the presence of chromatic change of magnification the chief ray is dispersed and intersects the image plane at the different heights.

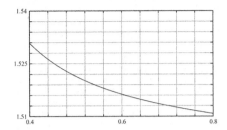

Figure 7.1 Change of index of refraction with wavelength, 0.4 to 0.8 μm, for glass Schott N-BK7.

Figure 7.2 Left, a marginal ray is dispersed at a single surface producing chromatic change of focus, and, right, a chief ray is dispersed producing chromatic change of magnification. Rays at the F, d, and C lines have been traced from the refracting surface to the ideal image plane.

Optical glass is characterized in the visible spectrum by its index, n_d, of refraction at the Fraunhofer d-line, $\lambda = 587.6$ nm, and by its v-number, $v = (n_d - 1)/(n_F - n_C)$, where n_F and n_C are the Fraunhofer F and C lines at $\lambda = 486.1$ nm and $\lambda = 656.2$ nm, respectively.

For a system of surfaces, the chromatic coefficients are given by,

$$\partial_\lambda W_{111} = \sum_{i=1}^{j} \left(\bar{A} \Delta \left(\frac{\partial n}{n} \right) y \right)_i, \tag{7.2}$$

and

$$\partial_\lambda W_{020} = \frac{1}{2} \sum_{i=1}^{j} \left(A \Delta \left(\frac{\partial n}{n} \right) y \right)_i, \tag{7.3}$$

where $\Delta \left(\frac{\partial n}{n} \right) = \frac{\partial n'}{n'} - \frac{\partial n}{n}$, $\partial n = n_F - n_C$, and $n = n_d$.

For a system of thin lenses, the chromatic coefficients are given by,

$$\partial_\lambda W_{111} = \sum_{i=1}^{j} \left(\frac{\phi}{v} y \bar{y} \right)_i, \tag{7.4}$$

and

$$\partial_\lambda W_{020} = \frac{1}{2} \sum_{i=1}^{j} \left(\frac{\phi}{v} y^2 \right)_i, \tag{7.5}$$

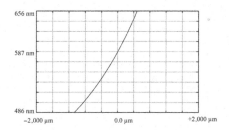

Figure 7.3 Chromatic focal shift curve for a singlet lens made out of BK7 glass.

where ϕ is the optical power of a thin lens. These second-order formulas are simple and often useful, as they describe the behavior of the chromatic change of focus and magnification. When the marginal ray height, y, at a thin lens is zero, there is no contribution to the chromatic aberrations by that lens, and, when the chief ray height, $\bar{y}$, is zero there is no contribution to the chromatic change of magnification by that lens.

For a thin lens with the stop aperture at the lens, the chromatic coefficients in terms of structural coefficients are,

$$\partial_\lambda W_{020} = \frac{1}{2} y_P^2 \phi \sigma_L, \tag{7.6}$$

and

$$\partial_\lambda W_{111} = 2 \mathcal{K} \sigma_T, \tag{7.7}$$

where the structural coefficients are $\sigma_L = \frac{1}{v}$ and $\sigma_T = 0$. The change of structural aberration coefficients with stop shifting are $\sigma_L^* = \sigma_L$ and $\sigma_T^* = \sigma_T + \bar{S}_\sigma \sigma_L$, where the structural stop shifting parameter is $\bar{S}_\sigma = \frac{y_P \bar{y}_P \phi}{2\mathcal{K}}$.

A singlet lens with a focal length of 100 mm made out of glass BK7 contributes chromatic change of focus, as shown in Figure 7.3.

The plot in Figure 7.3 shows the change in focal length in the horizontal axis vs. the wavelength in the vertical axis. For the primary wavelength which is the d-line the focal length is 100 mm. This plot is called the chromatic focal shift curve, and it is a useful plot to describe and understand chromatic aberration.

7.2 Optical Glass

Optical glass is manufactured by companies such as CDGM, Hoya, Ohara, Nikon, and Schott. Glass is made out of silica, soda-lime, and additives like lead oxide, Barium (BA), Boron (B), Fluorine (F), Lanthanum (LA), and Phosphorous (P).

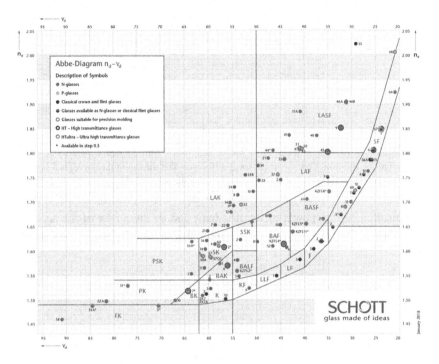

Figure 7.4 Abbe diagram as a function of the index of refraction and ν-number for the Schott catalog of glass families.
(Source: Schott)

Glasses are divided into crown glasses and flint glasses at a ν-number of approximately 50. Except for a few glasses, crown glasses have a ν-number larger than 50, and flint glasses less than 50. As silica is replaced with lead-oxide, the glass density and index of refraction increase, and the ν-number decreases. With the increase of lead-oxide, glass properties change, and several glass families are defined in the Schott catalog: K (crowns), KF (crown flints), LLF (very light flints), F (flints), and SF (dense flints). With additives, other families are also defined: FK (Fluorite crowns), PK (Phosphate crowns), BK (Boron crowns), BAK (Barium crowns), BAF (Barium flints), LAK (Lanthanum crowns), and LAF (Lanthanum flints). The glass families are shown in the Abbe diagram in Figure 7.4, which is a plot of glass index of refraction vs. the ν-number.

7.3 Thin Achromatic Doublet

Chromatic aberration can be corrected in a combination of two thin lenses with different ν-numbers and focal lengths. If $\phi = \phi_1 + \phi_2$ is the power of the lens

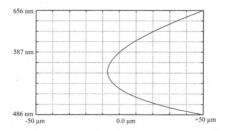

Figure 7.5 Chromatic focal curve for an achromatic thin doublet with a focal length of 100 mm. The maximum focal difference, this the sag of the curve, is 57 μm. Note that the extreme wavelengths focus at the same plane.

combination, ϕ_1 the power of the first lens, and ϕ_2 the power of the second lens, then to correct for chromatic aberrations in a thin doublet lens we must have,

$$\rho_1 = \frac{v_1}{v_1 - v_2}, \qquad (7.8)$$

and

$$\rho_2 = -\frac{v_2}{v_1 - v_2}, \qquad (7.9)$$

where $\rho_1 = \phi_1/\phi$ and $\rho_2 = \phi_2/\phi$. These equations result in the ratio of the optical powers needing to be equal to the negative of the ratio of the v-numbers,

$$\phi_1/\phi_2 = -v_1/v_2. \qquad (7.10)$$

The individual thin lens optical power is inversely proportional to the v-number difference. To minimize monochromatic aberration, it is important to minimize the optical power of the thin lens components, which requires using glasses with a large v-number difference. The chromatic focal shift for an achromatic doublet made out for BK7 and F2 glasses is shown in Figure 7.5. The focal length of the lens elements are 43.32 mm and -76.44 mm, respectively.

The two extreme wavelengths, λ_F and λ_C, focus at the same location, and all other wavelengths focus shorter. The primary wavelength, λ_d, per design has no focus error, and the maximum focal difference, this is the sag of the curve, is 57 μm. The residual chromatic aberration of Figure 7.5 is called the secondary spectrum and, for a third wavelength λ, it can be characterized by the glass partial dispersion ratio, defined by $P_\lambda = (n_\lambda - n_F)/(n_F - n_C)$, where n_λ is the index of refraction for a wavelength that might be brought to focus with the λ_F and λ_C wavelengths.

The chromatic change of focus is normally calculated for the λ_F and λ_C wavelengths. However, for λ_F and an intermediate wavelength, λ, the chromatic change of focus is,

$$\partial_\lambda W_{020} = \frac{1}{2} \sum_{i=1}^{j} \left(\phi \frac{P_\lambda}{v} y^2 \right)_i . \qquad (7.11)$$

The condition to make a thin achromatic doublet to provide the same focal length for three wavelengths is,

$$\partial_\lambda W_{020} = \frac{1}{2} \phi y^2 \left(\frac{P_1 - P_2}{v_1 - v_2} \right) = 0, \qquad (7.12)$$

which requires that the partial dispersion ratio of the glasses for the third wavelength be the same, $P_1 - P_2 = 0$.

For most glasses the partial dispersion ratio as a function of the v-number follows the line $P = a + bv$, where a and b are coefficients. Then the partial dispersion ratio difference for two glasses is $P_1 - P_2 = b(v_1 - v_2)$, and, therefore, there is no solution to bring a third wavelength to focus because the v-number difference must be finite for an achromat. Glasses that follow the line $P = a + bv$ are called normal glasses. Fluorite and phosphate glasses do not follow the normal glass line, and can be used to bring three wavelengths to a common focus. An example is a doublet made out of N-PSK57 and KZFS6 glasses from Schott, with element focal lengths of 29.078 mm and −41.0 mm to provide a focal length of 100 mm. The chromatic focal shift is shown in Figure 7.6. Note the reduction in the amount of secondary spectrum. When a lens system focuses three wavelengths to a common focal point, it is referred to as an apochromatic lens.

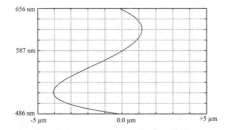

Figure 7.6 Chromatic focal shift curve for an apochromatic doublet made out of N-PSK57 and KZFS6 glasses. The v-number difference is $v_1 - v_2 = 68.39 - 48.51 = 19.88$.

7.4 Apochromatic Triplet

The secondary spectrum of a thin achromat doublet depends on the focal length and on the partial dispersion ratio difference of the glasses used. An apochromatic thin triplet lens can be made out of two thin achromatic doublets having a common glass. By the choice of glass and optical power of the individual doublets then it is possible to bring three wavelengths to a common focus.

A simple but powerful method to design an apochromatic objective lens is to plot the chromatic focal shift for several doublets made out of different glasses and with a focal length of 100 mm. If S_1 and S_2 are the sags of the chromatic focal shift curves for those doublets, then changing the focal length of one doublet by the ratio $\pm S_1/S_2$ would produce an apochromatic lens when both doublets are combined. The sign of the ratio depends on the sign difference of the sag of the chromatic focal shift curves.

For example, a first 100 mm focal length achromatic doublet made out of N-FK51 and N-KZFS4 glasses has a sag in the chromatic focal shift curve of -16.6 μm, and a second doublet made out of N-KZFS4 and N-SF15 glasses has a sag of -89 μm. By changing the focal length of the second doublet by a factor of $-89/16.6$ and then combining it with the first doublet, an apochromatic triplet is obtained. The chromatic focal shift curves are shown in Figure 7.7. Since the chromatic correction does not depend on the element shapes, the common glass lens elements can be combined into a single lens element. This procedure is particularly useful for finding apochromatic objectives that operate in the infrared or UV regions of the spectrum where the ν-number is not defined.

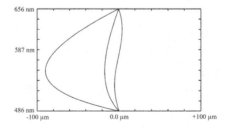

Figure 7.7 Chromatic focal shift curves. Left, for a N-KZFS4/N-SF15, $f' = 100$ mm thin doublet; middle, for a N-KZFS4/N-FK51, $f' = 100$ thin doublet; and, right, for a thin apochromatic triplet combining doublets with focal lengths of -536 mm and 100 mm. The focal shift curve for the apochromat triplet has been scaled up by a factor of 25.

Table 7.1 *Apochromatic and aplanatic objective,* f *= 1,000 mm, at* F*/6.7 0.425 μm to 0.75 μm*

Surface	Radius, mm	Thickness, mm	Glass
1	581.3493	10	N-KZFS11
2	167.8804	2	
3	170.2867	18	N-BAF4
4	328.8949	2	
5	290.9040	20	N-FK51A
6	−1,656.1579		

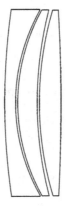

Figure 7.8 Apochromatic and aplanatic triplet lens objective.

The design of an achromatic or apochromatic lens objective also requires the control of spherical aberration and its dependence with color, called spherochromatism aberration, and also of the control of coma aberration. Thus, an optimized design involves a balance of aberrations. An example of an apochromatic and aplanatic objective is shown in Figure 7.8, and its constructional data is given in Table 7.1.

7.5 Glass Selection

Optical glass catalogs describe more than 100 different glass types, and then the question is how to select glasses. The selection depends on the application and on the glass optical, thermal, mechanical, and environmental properties. In addition to the index of refraction and v-number, glasses for example also differ in partial dispersion ratios, in optical transmission, in how the index of refraction changes with temperature, in coefficient of thermal expansion, durability, hardness, and cost.

For an achromatic doublet the glass choice is often driven by having a large v-number difference to minimize aberrations such as spherochromatism. For an apochromatic objective, fluorite glasses and short flint glasses which do not follow the normal glass line are used. Glasses with opto-thermal coefficients similar to aluminum, invar, or titanium may be chosen to athermalize a lens system. Flint glasses absorb more light toward the UV, and fused silica may be used instead of glass. The lanthanum glasses have a higher index of refraction that reduces lens curvatures and aberration, and can provide a useful v-number difference. Glass density might be important to reduce lens weight. BK7 glass is produced in large quantities, and serves as a useful reference, as the price of other glasses per pound is compared to that of BK7 glass. In addition, BK7 glass can be obtained in relatively large sizes.

7.6 Thermal Change of Focus and Magnification

Let us consider a thin lens where the change of focus, W_{020}, is measured with respect to the exit pupil plane, and given by,

$$W_{020} = \frac{y^2}{2} \phi = (n-1) \left(\frac{1}{r_1} - \frac{1}{r_2} \right) \frac{y^2}{2}, \tag{7.13}$$

where ϕ is the optical power of the thin lens.

The coefficient of thermal expansion, α, of the lens refracting material is,

$$\alpha = \frac{1}{r_1} \frac{\partial r_1}{\partial t} = \frac{1}{r_2} \frac{\partial r_2}{\partial t}. \tag{7.14}$$

Then, assuming that the change of index of refraction of air with temperature is negligible, the thermal change of focus is,

$$\partial_t W_{020} = \frac{y^2}{2} \phi \left(\frac{1}{n-1} \frac{\partial n}{\partial t} - \alpha \right) = \frac{y^2}{2} \phi \gamma, \tag{7.15}$$

where γ is the opto-thermal coefficient of the refracting material,

$$\gamma = \frac{1}{n-1} \frac{\partial n}{\partial t} - \alpha. \tag{7.16}$$

For a system of thin lenses, the thermal change of focus can be written as,

$$\partial_t W_{020} = \frac{1}{2} \sum_{i=1}^{n} \left(y^2 \phi \gamma \right)_i, \tag{7.17}$$

Table 7.2 *Opto-thermal coefficients*

Material	$\gamma[^1\!/_K] \times 10^{-6}$
LITHOSIL-Q	21.89
N-ZK7	9.24
N-BAK4	0.07
N-BK7	−1.21
N-F2	−2.36
N-FK5	−10.83
N-FK51A	−25.54

and the thermal change of magnification as,

$$\partial_t W_{111} = \sum_{i=1}^{n} (y\bar{y}\phi\gamma)_i. \tag{7.18}$$

These thermal aberration formulas are similar to the chromatic aberration formulas for a thin lens. The difference is that, instead of using the reciprocal of the glass v-number, the thermal aberration formulas use the opto-thermal coefficient that is in the order of $10^{-5}/K^0$. In comparison, the glass v-number may range from 20 to 90 and, thus, the change of focus due to temperature changes can be a small effect. The opto-thermal coefficient, γ, can be positive or negative, depending on the material, as shown in Table 7.2 for a few materials.

A lens system with a focal length, or other property, that is insensitive to changes in temperature is called an athermal system. To have an achromatic and athermal doublet, the additional requirement is,

$$\partial_t W_{020} = \frac{y^2}{2}(\phi_1\gamma_1 + \phi_2\gamma_2) = \frac{y^2}{2}\frac{\phi}{v_1 - v_2}(v_1\gamma_1 - v_2\gamma_2) = 0, \tag{7.19}$$

or,

$$v_1\gamma_1 = v_2\gamma_2. \tag{7.20}$$

The equivalent opto-thermal coefficient for an achromatic doublet is,

$$\gamma = \frac{v_1\gamma_1 - v_2\gamma_2}{v_1 - v_2}. \tag{7.21}$$

In the same way that an apochromatic triplet can be designed out of two achromatic doublets sharing a common glass, an achromatic doublet that is athermal can be designed. If γ_A is the equivalent opto-thermal of an achromatic thin doublet, and γ_B is that of another doublet, then the ratio of the focal lengths must be,

$$\frac{f_A}{f_B} = -\frac{\gamma_A}{\gamma_B},$$

(7.22)

for the combination of both doublets to be athermal. The thermal change of focal length can be weak, so it may not be necessary to exactly match the ratio of the focal lengths to the ratio of the opto-thermal coefficients.

As an example, an achromatic and athermal triplet can be made out of thin lenses with glasses N-BK7, N-F2, and N-ZK7 with focal lengths of 42.72 mm, −76.24 mm, and −3,478.78 mm, respectively to yield a triplet lens with a focal length of 100 mm. The combination of N-BK7 and N-F2 glass gives a nearly achromatic and athermal doublet. The use of the N-ZK7 glass lens is to help tune the aberrations to zero.

Because both the chromatic change of focus and the thermal change of focus depend on the square of the aperture, it is best to correct for these aberrations at locations in a lens system where the marginal ray height is maximum. The thermal change of focus formula is based on a linear model and, for large temperature ranges, the model may be in significant error.

7.7 Techniques for Correcting Chromatic Aberration

A favorite strategy to correct aberration in a lens system is to first do the monochromatic aberration correction, and then do the chromatic aberration correction. This separates the tasks and simplifies the design process.

The correction of chromatic aberration in a multi-lens system can be done by making every lens an achromatic doublet or triplet. However, this procedure adds cost and is often redundant.

To correct for the chromatic aberrations, two effective degrees of freedom are needed. These can be two separated dispersive interfaces, or one dispersive interface and its position in the lens system. The symmetry of the system about the aperture stop may be used to correct the chromatic change of magnification since it is an odd aberration. If the system is optically fast, more dispersive interfaces can be added to avoid very strong surface curvatures.

A first technique is to design a lens monochromatically using a single glass type with an index of refraction near $n_d = 1.65$. Near this index there are several glasses that have different v-numbers; for example, LAK21, LAK7, LAK22, LAK5, BAF61, KZFSN5, and SF2 in the Schott glass catalog. Once the monochromatic design is done, the glass of positive lenses is chosen to be a crown, and the glass of negative lenses is chosen to be a flint. Since the index for the nominal wavelength is roughly the same, the monochromatic correction will not change significantly. However, the chromatic aberrations will change, and sometimes they can be corrected by a proper selection of the glass v-number for each lens element.

An optical surface, called a buried surface, is surrounded by two glasses with the same index of refraction, $n_d = n_d'$, but with different v-numbers, $v \neq v'$. Thus, the rays of light for the λ_d wavelength do not refract at the buried surface, but rays at other wavelengths do refract at the buried surface, as there is an index difference for those wavelengths. The radius of curvature of the buried surface and the v-number difference serve as effective variables to control chromatic aberration. The buried surface provides a degree of freedom to control chromatic aberration.

A second technique is, again, to design the lens monochromatic first, and then identify the position for the stop aperture that would correct the chromatic change of magnification aberration. If that position coincides with a lens element, then that lens can be split by a buried surface to correct for the chromatic change of focus aberration. At this point both chromatic aberrations would be corrected. Then the aperture stop can be returned to its original position, where it is best for the monochromatic correction. The chromatic correction will remain as stop shifting, $\sigma_L^* = \sigma_L$, and $\sigma_T^* = \sigma_T + \bar{S}_\sigma \sigma_L$ will not change the chromatic change of magnification, since there is no chromatic change of focus aberration, $\sigma_L = 0$, and no chromatic change of magnification, $\sigma_T = 0$.

As an example, consider the monochromatic quartet from the lens design problem of the 1990 International Lens Design Conference, as shown in Figure 7.9 (top). The chromatic aberration for the on-axis field position and for a field of 15° is shown in Figure 7.10 (top). To achromatize this lens system, a buried surface was included in the third lens at a point where the

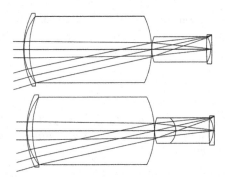

Figure 7.9 Top, monochromatic quartet; bottom, a buried surface was included in the third lens to correct for chromatic aberration.

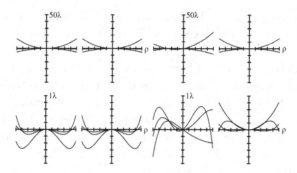

Figure 7.10 Top, wave fans for the monochromatic quartet with a scale of 50 waves showing both chromatic change of focus and chromatic change of magnification aberration. Bottom, wave fans for the chromatically corrected lens with a scale of one wave. The wave fans are for the field positions 0° and 15° and for the λ_F, λ_d, and λ_C wavelengths.

chromatic change of magnification was canceled, and the radius of curvature of the buried surface and the ν-number difference was used to correct for chromatic change of focus. Figure 7.9 (bottom) shows the chromatically corrected lens, and Figure 7.10 (bottom) shows the aberration correction residual.

A third technique to correct for chromatic change of focus and chromatic change of magnification is to introduce two buried surfaces in two different lens elements. This technique is used in the double Gauss lens that is discussed in Chapter 12.

7.8 Diffractive Optical Elements

Diffractive Optical Elements (DOE) change the path of light through the phenomenon of diffraction. In particular, they can deviate light, converge or diverge light, and introduce or correct aberration. Consider the grating equation,

$$n' \sin (I') - n \sin (I) = \frac{m\lambda}{d}, \qquad (7.23)$$

where m is the diffractive order, d is the grating period, and I and I' are the angles of incidence and diffraction. For the zero-order, $m = 0$, the grating equation becomes Snell's law.

Figure 7.11 (top) shows an amplitude grating where 50% of the light is lost and 50% is diffracted into multiple orders, and where light modulation is accomplished by a periodic variation of the light field amplitude. Figure 7.11

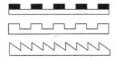

Figure 7.11 Top, an amplitude grating; middle, a phase grating; bottom, a phase and blazed grating.

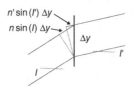

Figure 7.12 Two close and parallel meridional rays are incident in a plane grating. The angle of incidence is I, and the angle of diffraction is I'.

(middle) shows a phase grating where 100% of the light is diffracted into multiple orders and where modulation is accomplished by a periodic variation of the light field phase. Figure 7.11 (bottom) shows a phase and blazed grating where 100% of the light is diffracted into a single order and light modulation is accomplished by a periodic variation of the light field phase. The blaze of a grating refers to the profile of the grating grooves and determines how light is distributed among the diffracted orders.

Figure 7.12 shows two close and parallel meridional rays separated by a distance, Δy, incident in a plane grating. The optical path difference, or phase difference, $\Delta\phi$, between them after diffraction is,

$$\Delta\phi = (n' \sin (I') - n \sin (I))\Delta y = \frac{m\lambda}{d} \Delta y. \tag{7.24}$$

In the limit of small Δy, we obtain the derivative of the phase with respect to the ray intersection coordinate, y,

$$\frac{\partial\phi}{\partial y} = (n' \sin (I') - n \sin (I)) = \frac{m\lambda}{d}. \tag{7.25}$$

This relationship represents a ray tracing equation whenever the phase function, $\phi(y)$, is known. A first conclusion is that a grating introduces a linear phase change as a function of position, and deviates light. Then a grating that introduces a quadratic phase would make light converge or diverge, and this is a diffractive lens, as shown in Figure 7.13. A second conclusion is that the grating geometry, when it is thought of as an interferogram, introduces a wavefront deformation that corresponds to that represented by the interferogram.

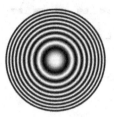

Figure 7.13 An amplitude diffractive lens.

The ray deviation of a grating can be written to first-order as,

$$\delta = n'i' - ni = \frac{m\lambda_d}{d}.$$ (7.26)

The grating dispersion can be written as,

$$\Delta = \delta_F - \delta_C = \frac{m(\lambda_F - \lambda_C)}{d}.$$ (7.27)

Therefore, the v-number that corresponds to a grating or diffractive lens is,

$$\nu_{diffractive} = \frac{\delta}{\Delta} = \frac{\lambda_d}{\lambda_F - \lambda_C} \cong -3.5.$$ (7.28)

Two important features are: (1) diffractive optical elements are strongly dispersive, much more than glass, and (2) there is more diffraction for longer wavelengths than for short wavelengths. This is the opposite of refraction, where there is more refraction for shorter wavelengths than for longer wavelengths.

The chromatic change of focus of the combination of a refractive thin lens and a diffractive lens in contact is,

$$\partial_\lambda W_{020} = \frac{1}{2}\left(\frac{\phi_{refractive}}{\nu_{refarctive}} + \frac{\phi_{diffractive}}{\nu_{diffractive}}\right)y^2.$$ (7.29)

Because of the negative v-number of diffractive optical elements and their strong dispersion, they can be used to correct for chromatic aberration using elements with positive optical power. However, the amount of optical flux directed to a given diffractive order depends also on the wavelength. The amount of light that is not directed to the intended order becomes stray light and decreases image contrast. This is a major problem of DOEs in broadband applications.

Lens design programs model diffractive optical elements by specifying a phase function. One example of an axially symmetric phase function is,

Table 7.3 *Prescription for an achromatic lens using the Sweatt model of a DOE.* $f' = 100$ mm

Surface	Radius	Thickness	Glass
1	54.6098	5	BK7
2	Plano	0	$n_F = 4,861.3$
			$n_d = 5,875.6$
			$n_C = 6,562.7$
3	-1.013×10^7	96.565	

$$\varphi = A\frac{2\pi}{\lambda}\left(x^2 + y^2\right) + B\frac{2\pi}{\lambda}\left(x^2 + y^2\right)^2 + C\frac{2\pi}{\lambda}\left(x^2 + y^2\right)^3 \ldots, \qquad (7.30)$$

where A, B, and C are coefficients that define the diffractive optical element. The derivative of the phase function and the shape of the substrate on which the diffractive structure is made are used to calculate the direction of the diffracted ray. The second-order term in the phase function introduces focus, this is optical power. The fourth-order and higher-order terms allow us to introduce spherical aberration.

By multiplying the grating equation by,

$$\frac{n'\cos\left(I'\right) - n\cos\left(I\right)}{n'\cos\left(I'\right) - n\cos\left(I\right)} = 1, \qquad (7.31)$$

and defining,

$$\tan\left(\alpha\right) = \frac{1}{n'\cos\left(I'\right) - n\cos\left(I\right)}\frac{m\lambda}{d}, \qquad (7.32)$$

the grating equation can be re-arranged as

$$n'\sin\left(I' - \alpha\right) = n\sin\left(I - \alpha\right). \qquad (7.33)$$

When the index of refraction is large, $n = 10{,}000$, then α becomes negligible and the grating equation becomes nearly Snell's law. The implication is that light diffraction of a ray can be modeled by refraction. Thus, for a given optical power, a plano convex lens with a very high index of refraction becomes almost like a parallel plate of glass and models a diffractive lens of the same optical power. This is known as the Sweatt model, and it is a simple way to model a diffractive optical element. To model light dispersion, the index of refraction can be set to be $n = \lambda \cdot 10^4$ using the Schott dispersion formula. Table 7.3 provides the prescription of an achromatic singlet lens that has a diffractive rear surface, as modeled by the Sweatt model.

Further Reading

de Albuquerque, Bráulio Fonseca Carneiro, Sasián, José, de Sousa, Fabiano Luis, Montes, Amauri Silva. "Method of glass selection for color correction in optical system design," *Optics Express*, 20 (2012), 13592–611.

Hartmann, P. *Optical Glass*, Vol. PM249 (Bellingham, WA: SPIE Press, 2014).

Li, C. L., Sasián, J. "Adaptive dispersion formula for index interpolation and chromatic aberration correction," *Optics Express*, 22(1) (2014), 1193–202.

Mercado, R. I. "Design of apochromats and superachromats," *Proceedings of SPIE*, 1 (1992), 270–96.

O'Shea, D., Suleski, T., Kathman, A., Prather, D. *Diffractive Optics: Design, Fabrication, and Test* (Bellingham, WA: SPIE Press Book Vol. TT62, 2003).

O'Shea, Donald C. "Monochromatic quartet: a search for the global optimum", Proceedings of SPIE 1354, 1990 International Lens Design Conference (1990); doi: 10.1117/12.47896.

Reshidko, D., Sasián, J. "Method of calculation and tables of opto-thermal coefficients and thermal diffusivities for glass," *Proceedings of SPIE*, 8844 (2013), doi: 10.1117/12.2036112.

Reshidko, Dmitry, Nakano, Masatsugu, Sasián, José. "Ray tracing methods for correcting chromatic aberrations in imaging systems," *International Journal of Optics*, 2014 (2014), Article ID 351584.

Reshidko, Dmitry, Sasián, José. "Algorithms and examples for chromatic aberration correction and athermalization of complex imaging systems," Proceedings of SPIE 9293, International Optical Design Conference 2014, 92931K (2014).

Sasián, José, Gao, Weichuan, Yan, Yufeng. "Method to design apochromat and super-achromat objectives," *Optical Engineering*, 56(10) (2017), 105106.

Sweatt, William C. "Describing holographic optical elements as lenses," *Journal of the Optical Society of America*, 67 (1977), 803–8.

Links to optical glass catalogs:
www.us.schott.com,
www.oharacorp.com,
www.hoyaoptics.com,
cdgmglass.com,
www.nikon.com/products/glass/lineup/materials/optical/catalog/index.htm.

8

Combinations of Achromatic Doublets

The achromatic doublet is a fundamental building block in lens design because it is corrected for chromatic aberrations, and can also be corrected for spherical aberration and coma aberration. The early lens designers explored all combinations of two achromatic doublets. This chapter discusses some of the solutions found by those designers. In doing so, insight is gained into how simple lens combinations are designed. Emphasis is given to how the primary aberrations are controlled in doublet combinations, as this knowledge is important to become skilled in lens design. Providing degrees of freedom to correct the primary aberrations is a first step toward the optimization of a lens. In practice, the primary aberrations may not be fully corrected so that higher order aberrations might be balanced against the primary aberration residuals. Once a primary aberration solution was reached in the examples given in this chapter, then they were optimized with real rays in a lens design program by minimizing RMS spot size across the field of view. Thus, a lens design method is to find a primary aberration solution and then optimize it with real ray tracing.

8.1 Structural Aberration Coefficients of a Thin Achromatic Doublet

Table 8.1 provides the structural aberration coefficients for a thin achromatic doublet with the stop aperture at the doublet. The surfaces are assumed to be spherical in shape, and the ratios of the element optical power to the total power are $\rho_1 = \phi_1/\phi$ and $\rho_2 = \phi_2/\phi$.

In Table 8.1, σ_L and σ_T are the structural aberration coefficients for the chromatic change of focus, $\partial_\lambda W_{020} = \frac{1}{2} y_P^2 \phi \sigma_L$, and the chromatic change of magnification, $\partial_\lambda W_{111} = 2\mathcal{H}\sigma_T$. If $Y = \frac{1+m}{1-m}$ is the conjugate factor at which the doublet works, then the individual conjugate factors for the elements are,

81

Table 8.1 *Structural aberration coefficients of a thin doublet*
Stop aperture at the doublet

$$\sigma_L = \frac{\rho_1}{\nu_1} + \frac{\rho_2}{\nu_2}$$

$$\sigma_T = 0$$

$$\sigma_I = \rho_1{}^3\left(A_1 X_1{}^2 - B_1 X_1 Y_1 + C_1 Y_1{}^2 + D_1\right) + \rho_2{}^3\left(A_2 X_2{}^2 - B_2 X_2 Y_2 + C_2 Y_2{}^2 + D_2\right)$$

$$\sigma_{II} = \rho_1{}^2(E_1 X_1 - F_1 Y_1) + \rho_2{}^2(E_2 X_2 - F_2 Y_2)$$

$$\sigma_{III} = \rho_1 + \rho_2 = 1$$

$$\sigma_{IV} = \frac{\rho_1}{n_1} + \frac{\rho_2}{n_2}$$

$$\sigma_V = 0$$

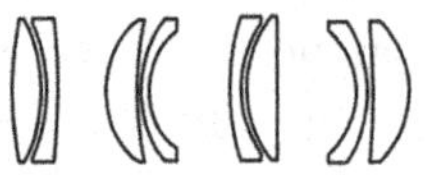

Figure 8.1 Left, crown glass in front solutions, and, right, flint glass in front solutions shown as thick lenses.

$$Y_1 = \frac{Y - \rho_2}{\rho_1}, \tag{8.1}$$

and

$$Y_2 = \frac{Y - \rho_1}{\rho_2}. \tag{8.2}$$

For a given conjugate factor, Y, and as a function of the shape factors, X_1 and X_2, the plot of spherical aberration is a hyperbola. Thus, there might be two solutions that produce a given amount of spherical aberration, σ_I.

The plot of coma as a function of the shape factors X_1 and X_2 is a straight line. Thus, there might be a solution that produces a given amount of coma aberration, σ_{II}.

Astigmatism aberration, σ_{III}, is fixed when the stop is at the doublet, as well as Petzval field curvature, σ_{IV}. There is no chromatic change of magnification, σ_L, or distortion aberration, σ_V.

Depending on whether the crown glass or the flint glass is in front, there are four solutions for a thin achromatic doublet that is aplanatic, as shown in Figure 8.1.

The fact that there are several solutions is important, because often one solution performs best for a given application. In multi-lens systems the individual doublet solutions give place to a plurality of lens system solutions when these solutions are interchanged.

If there is no requirement for the control of coma aberration, then a doublet lens can have the same radius of curvature but with opposite sign in two contiguous surfaces that can be cemented to form a single lens. Lenses with a diameter of less than 50 mm can be cemented with UV curing cement. Large lenses, if cemented, may experience stress, which can alter the shape of the lenses or introduce birefringence. For cemented doubles with strong curvatures it is a good practice to avoid having ray total internal reflection if, for any reason, air gets between the cement and the glass.

8.2 Field Curvature of a Thin Achromatic Doublet

The Petzval field curvature of a thin achromatic doublet in air is given by,

$$C_{Petzval} = \frac{1}{\rho_{Petzval}} = -\phi \cdot \sigma_{IV} = -\left(\frac{\phi_1}{n_1} + \frac{\phi_2}{n_2}\right) = -\frac{\phi}{v_1 - v_2}\left(\frac{v_1}{n_1} - \frac{v_2}{n_2}\right). \quad (8.3)$$

For a doublet with a focal length of 100 mm made of N-BK7 and N-F2 glasses, the element focal lengths are 43.26 mm and -76.25 mm, respectively, and the Petzval radius is, $\rho_{Petzval} = -139.99$ mm. In comparison, a singlet lens with a focal length of 100 mm and made out of N-BK7 glass has a Petzval radius of -151.7 mm, which is longer than that of the doublet. Note that an achromatic doublet with a negative focal length of -100 mm and using the same glasses, has a Petzval radius of $+139.99$ mm.

With the advent of new glasses manufactured by Otto Schott and Ernst Abbe, by 1886 a longer Petzval radius in a doublet was possible. The index of refraction of the crown element is maximized to reduce its contribution to the Petzval sum, and the index of refraction of the flint element is minimized to increase its contribution, but opposite in sign, to the Petzval sum. For example, with glasses N-BAK1 and N-LLF6, a Petzval radius of -185 mm is achieved for a 100 mm focal length doublet lens. With the advent of the lanthanum glasses in 1930, even a longer Petzval radius is possible. Using glasses N-LAK34 and N-F2, a Petzval radius of -200 mm is obtained with a v-number difference of 18.06. As the v-number difference of the glasses used decreases, the Petzval radius also decreases. In addition, the lens curvatures increase, and the amount of aberration contributed by the surfaces also increases. The achromatic doublet made with glasses prior to the new glasses developed by Schott and Abbe is called an old achromat. The achromat doublet made with the new glasses and with an increased Petzval radius is called the new achromat. Figure 8.2 compares the singlet lens, the old achromatic doublet, and the new achromat doublet. In a new achromatic cemented

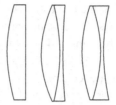

Figure 8.2 A singlet lens, an old achromatic doublet, and a new achromatic doublet. Note that the new achromat doublet has a decreased thickness variation across its aperture; this explains its longer Petzval radius.

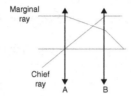

Figure 8.3 First-order layout of a two component, A and B, microscope objective. The chief ray and the marginal ray are drawn.

doublet, there may not be a solution for correcting spherical aberration, and it may suffer from a strong amount of positive spherical aberration.

8.3 Lister Microscope Objective

Joseph Lister found out empirically that, by the proper placement of two achromatic doublets, coma aberration could be corrected in a microscope objective. The chromatic aberrations and spherical aberration can be independently corrected at each doublet. The coma aberration of one doublet can be corrected by the other doublet, thus forming an achromatic and aplanatic objective.

The use of structural aberration coefficients is illustrated with the following example. The achromatic doublets have optical powers, ϕ_A and ϕ_B, and can be individually corrected for spherical aberration so that, $\sigma_{IA} = \sigma_{IB} = 0$. A first-order model of an objective is illustrated in Figure 8.3. The system is normalized by setting $\mathcal{K} = 1$, the total power is equal to one, $\phi = 1$, and the marginal ray height at the first doublet is also equal to one, $y_A = 1$. The design is done in reverse, with the object at infinity. The aperture stop is at the first doublet, $\bar{y}_A = 0$, and telecentricity is required in image space, which results in the following optical powers; $\phi_B = 1$, $\phi_A = 1 - y_B$, and $\bar{y}_B = 1$.

Using the structural coefficients for a system, see Table 7 in Appendix 3, we can write for coma aberration of the combination,

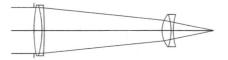

Figure 8.4 Microscope objective with a focal length of $f = 5$ mm and an optical speed of $F/2.5$.

$$\sigma_{II} = \phi_A^2 y_A^2 \sigma_{IIA} + \phi_B^2 y_B^2 \sigma_{IIB}, \qquad (8.4)$$

where we have used the fact there is no spherical aberration from each of the components, and in particular, $\sigma_{IB} = 0$. Then a relationship between the marginal ray height, y_B, at the second component and the structural coefficients of each component results,

$$\sigma_{IIA} = -\frac{y_B^2}{(1 - y_B)^2} \sigma_{IIB}. \qquad (8.5)$$

If the marginal ray height is chosen to be $y_B = \frac{1}{2}$, then the structural aberration coefficients must be $\sigma_{IIA} = -\sigma_{IIB}$.

Figure 8.4 shows a solution using thick and cemented doublets. If the doublets are not cemented then there is more flexibility for correcting aberration. Requiring that the individual cemented doublets be corrected for spherical aberration locks their optical shape, and the correction of coma aberration for the combination depends on the doublet separation. This produces a near telecentric solution, as is shown in Figure 8.4. The doublets are made out of BK7 and F2 glasses and have focal lengths, $f_{1'} = 10$ mm and $f_{2'} = 5$ mm, respectively.

Because the stop aperture has been set in contact with the first doublet, the separation between doublets is relatively large. This reduces the unvignetted semi-field of view to $2°$. When the stop aperture is at each of the doublets, they contribute positive astigmatism. However, the stop aperture for the second doublet is remote and, because it contributes negative coma, it can contribute negative astigmatism. Stop shifting formulas indicate that the astigmatism of the doublet combination can be corrected whenever,

$$\sigma_{III} = (1 - y_B) + (1 + y_B \sigma_{IIB}) = 0. \qquad (8.6)$$

A more practical solution for a microscope objective is shown in Figure 8.5, where the doublet lenses, except for the diameter, are identical in construction and are individually corrected for spherical aberration for an object at infinity. The doublets separation is chosen to correct for coma aberration, and results in little astigmatism. There is spherical aberration, but it is insignificant since the

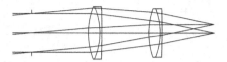

Figure 8.5 Microscope objective with a focal length of $f' = 5$ mm. The doublets are made out of BK7 and F2 glasses, have the same radii of curvature, and have a focal length of $f_{1'} = f_{2'} = 8.3$ mm.

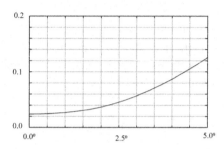

Figure 8.6 RMS wavefront error across the field of view in waves units. Image quality is limited by astigmatism aberration.

scale of the objective is small; it has a focal length of $f' = 5$ mm. There is also field curvature aberration, which can be tolerated and mitigated, by refocusing the objective. Distortion aberration is 1%. The RMS of the wavefront error across a semi-field of view of 5° is shown in Figure 8.6. The objective is diffraction limited over a semi-field of view of 3° at an optical speed of $F/2.5$. The numerical aperture is $NA = 0.2$.

The stop aperture has been located in front of the doublets to make the objective telecentric in image space. The doublet separation is not large, and an unvignetted semi-field of 5° with good image quality is achieved.

The resolving power of a lens, this is the ability to discern detail, depends on the numerical aperture. An increase in numerical aperture can be obtained by adding a lens that operates on the aplanatic-concentric principle. The first surface of the added lens satisfies, $\Delta(^u/_n) = 0$, and, therefore, it does not introduce spherical, coma, or astigmatism aberrations. It, however, increases the numerical aperture of the beams by a factor equal to the index of refraction of the lens. The second surface is concentric to the on-axis beam, so that there is no refraction of the marginal ray, $A = 0$, and, therefore, no spherical or coma aberrations are introduced. Figure 8.7 shows application of the aplanatic–concentric principle to increase the numerical aperture of a two-doublet microscope objective by adding a third singlet lens.

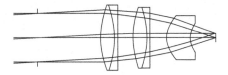

Figure 8.7 A third singlet lens increases the numerical aperture, from 0.2 to 0.3, of a two doublet microscope objective by a factor equal to the index of refraction of the singlet lens.

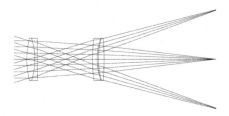

Figure 8.8 Petzval portrait objective. $f' = 144$ mm; $F/3.7$; FOV $= \pm 16.5°$.

8.4 Petzval Portrait Objective

After the public disclosure of the Daguerrotype process in 1839 there was a need for objective lenses that could provide sharp images at a fast optical speed. Daguerre's cameras were equipped with an achromatic landscape lens made by Chevalier in France and operated at speeds of about $F/16$. It took approximately half an hour to make a photographic exposure. Joseph Petzval tackled the problem and designed the lens shown in Figure 8.8, which made portrait photography a practical reality.

Petzval used two doublet lenses, individually corrected for chromatic aberrations and spherical aberration. The first doublet was cemented and contributed positive coma and positive astigmatism. The rim of the first doublet served as the aperture stop. The second doublet was split as to allow the control of spherical aberration and coma aberration, and had the flint element in front. Because of the negative coma contributed by the second doublet, and since the stop aperture was remote, enough negative astigmatism was contributed to artificially flatten the field of view. Astigmatism aberration contributed by the second doublet with a remote stop aperture is given by,

$$S_{III}^* = S_{III} + 2 \cdot \bar{S} S_{II} + \bar{S}^2 S_I. \tag{8.7}$$

Spherical aberration is zero, $S_I = 0$, and, with an increasing separation from the stop, the stop shifting parameter, $\bar{S}$, and negative astigmatism, $2 \cdot \bar{S} S_{II}$, increase, overcoming positive astigmatism, S_{III}. The amount of negative astigmatism that

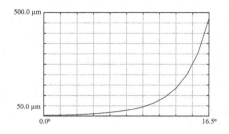

Figure 8.9 RMS spot size of the Petzval portrait objective.

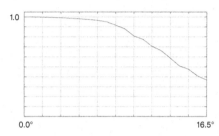

Figure 8.10 Relative illumination of the Petzval portrait objective lens.

Petzval chose to artificially flatten the field of view was similar to that of the Chevalier lens. This was to minimize the rms spot size over the field of view by satisfying the relationship, $W_{222} = -\frac{4}{5}W_{220P}$. The optical power of the doublets was chosen to maximize the optical power of the combination, and to minimize light vignetting by the second doublet. The first Peztval portrait lens had a focal length of approximately 144 mm, worked at $F/3.7$, and had a semi-field of view of $\pm16.5°$. The focal lengths of the doublets were approximately 200 mm and 300 mm, respectively. Since Petzval lenses are aplanatic, they have excellent resolving power near the field center. Because of the negative astigmatism and positive field curvature, image sharpness decreases towards the field edge. Figure 8.9 shows the geometrical RMS spot size as a function of the field of view. This indicates excellent image quality at the field center when the grain of the photographic film is taken to be 25 μm.

The increase in optical power of the Petzval lens, while controlling aberration, comes from splitting the task between two positive achromatic doublet lenses. This, in turn, makes the lens less sensitive to manufacturing errors as compared to other lenses that combine doublets with positive and negative optical power. In his drawing for the mechanical lens barrel, Petzval provided a generous light hood to avoid light from outside the field of view to enter the lens. He was aware of Fresnel reflections produced by bare lens surfaces, and the number of air-to-glass interfaces was a design concern. Figure 8.10 shows the relative illumination of the Petzval lens, including the effect of light vignetting.

8.5 Rapid Rectilinear Lens

There was a need for a lens that could provide a wider field of view than the Petzval portrait lens, and that could be faster than the Wollanston landscape lens. Several symmetrical doublet lens combinations were tried with crown-in-front and flint-in-front glasses. However, because, to artificially flatten the field, the doublets needed to be menisci, which conflicted with the correction of spherical aberration, the image quality was not satisfactory.

In 1866, J. H. Dallmeyer and H. A. Steinheil independently came up with a solution that became known as the rapid rectilinear lens. By decreasing the v-number difference of the glasses used, it was possible to control the spherical aberration of a symmetrical combination of two doublets that artificially flattened the field. A design with N-SSK5 and N-BK7 glasses is shown in Figure 8.11. The semi-field of view is 24°, the F-number is $F/8$, the focal length is 100 mm, and the Petzval radius is -121.2 mm. Figure 8.12 shows the

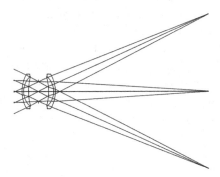

Figure 8.11 Rapid rectilinear lens with N-SSK5 and N-BK7 glasses. $f' = 100$ mm, FOV = ±24°, $F/8$.

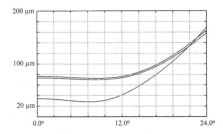

Figure 8.12 RMS spot size across the field of view. Bottom curve, rapid rectilinear; middle curve, flint-in-front with N-F2 and N-BK7 glasses; and upper curve, crown-in-front with N-BK7 and N-F2 glasses.

RMS spot size for the rapid rectilinear, and for the crown-in-front, and flint-in-front doublet solutions when N-BK7 and N-F2 glasses are used. The rapid rectilinear is a symmetrical lens about the stop aperture.

The spherical aberration of the rapid rectilinear depends on the ν-number difference, and it is possible to have this aberration positive, zero, or negative. However, because the doublets are achromatic and the element power is inversely proportional to the ν-number difference, the Petzval radius changes proportionally with the ν-number difference. For a design with N-SSK5 and N-BK7 glasses and a focal length of 100 mm, the Petzval radius is −122.2 mm. For a design with N-BK7 and N-F2 glasses and suffering from spherical aberration, the Petzval radius is −147.9 mm.

8.6 Concentric Lens

Because lens design depends on the glass optical properties, lens designers are attentive to new optical materials that can provide additional or enhanced degrees of freedom to correct aberrations. With the advent of the new optical glasses, in 1889 H. L. Schroder patented (U.S. Patent 404,506) a symmetrical lens, as shown in Figure 8.13, that provided a nearly flat field and excellent imaging at a speed of $F/16$ and a field of view of ±30°. Coma aberration and distortion were corrected by symmetry, astigmatism was controlled by the doublet separation from the stop aperture, and the chromatic aberrations were

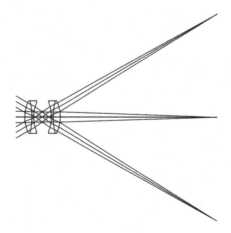

Figure 8.13 Schroder concentric lens using the new glasses developed by Schott and Abbe. For a focal length of 100 mm, the Petzval radius is −333 mm. FOV = ±30° at $F/16$.

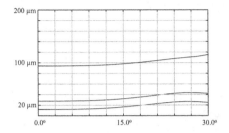

Figure 8.14 RMS spot size of the Schroder lens ($f' = 100$ mm) for speed $F/8$ top curve, $F/12$ middle curve, and $F/16$ lower curve.

individually corrected at each doublet. For a focal length of 100 mm and using modern glasses N-BAK1 and N-LLF6, the Petzval radius is -333 mm. However, spherical aberration was not corrected, but mitigated by working at $F/12$–$F/16$, becoming significantly objectionable at a speed of $F/8$, because spherical aberration grows with the fourth power of the aperture. Figure 8.14 shows the RMS spot size as a function of the field of view and the optical speed. Schroder's design had the exterior surfaces of each doublet nearly concentric, and was the first symmetrical objective using doublets to reduce the Petzval sum with the new glasses.

Given that the Petzval radius is longer than in objectives using the old glasses, little negative astigmatism needs to be introduced to arrive at an optimized design by artificially flattening the field of view.

8.7 Anastigmatic Lens

A second objective lens that took advantage of the new glasses was the anastigmatic lens of P. Rudolph (U.S. Patent 444,714). Instead of artificially flattening the field of view, Rudolph corrected the astigmatism by canceling the contribution of an old achromatic doublet with the opposite contribution of a new achromatic doublet. The merit, in part, was that the Petzval sum was nearly corrected, thus providing a substantially flat field. In addition, the positive spherical aberration of the new achromatic doublet was corrected with a negative contribution from the old achromatic doublet. The chromatic aberrations and coma aberration were also controlled. The lens is not symmetric about the aperture stop, and there are no degrees of design freedom left to correct for fourth-order distortion. However, it turned out that higher order distortion tended to balance the fourth-order, yielding a small net amount of distortion aberration. The second embodiment in Rudolph's patent optimized with modern glasses is shown in Figure 8.15. The front achromatic doublet

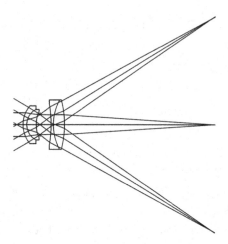

Figure 8.15 Anastigmatic lens using F2 and BK7 glasses for the front old achromatic doublet and using N-KF9 and N-SK16 glasses for the rear new achromatic
doublet. $f' = 100$ mm, $F/8$, FOV = $\pm 30°$.

uses old glasses, and it is designed as a thick meniscus lens, and the rear lens is
a new achromatic doublet. The focal length of the doublets are 262 mm and
128 mm, respectively, to provide a combination with a focal length of 100 mm.
The front doublet carries about half the optical power of the rear doublet.

A useful way to explain the anastigmatic lens is to realize that the new
achromatic doublet contributes a reduced amount of positive Petzval field
curvature aberration, and that the thick meniscus doublet provides minimal,
or negative Petzval field curvature to effectively help control the Petzval sum.
The separation from the stop aperture of the new achromat permits controlling
astigmatism. The use of a thick meniscus lens is one of the techniques in lens
design to control field curvature aberration.

Rudolph's lens became known as the anastigmatic lens. It appears that
Rudolph, in his patent, coined the term "anastigmatic," which means absence
of astigmatism for at least an off-axis field point. Thus, the term anastigmatic is
in the context of a lens that is substantially corrected for field curvature.
Figure 8.16 shows the field curves of the anastigmatic lens. Fourth-order
astigmatism balances higher order astigmatism, and there is a field point where
astigmatism is absent. The residual astigmatism shown in the field curves is
relatively small when it is compared to lenses that artificially flatten the field.
Distortion aberration is well controlled to less than 1.0%. The RMS spot size
across the field of view is shown in Figure 8.17, which represents a substantial
improvement over the rapid rectilinear lens. The Petzval radius is -257 mm
for a focal length of 100 mm, giving a ratio of about -2.6.

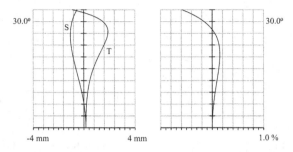

Figure 8.16 Field curves of the anastigmatic lens. $f' = 100$ mm, $F/8$, FOV = $\pm 30°$.

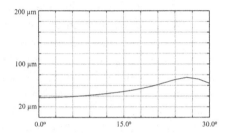

Figure 8.17 RMS spot size vs. field of view of the anastigmatic lens. $f' = 100$ mm, $F/8$, FOV = $\pm 30°$.

8.8 Telephoto Lens

A telephoto lens has a focal length that is longer that the length of the lens. The length, also known as the total track length (TTL), is measured from the first surface to the image plane, along the optical axis. The ratio of the focal length to the lens length is known as the telephoto ratio, and it is less than one. To achieve this property, a positive and a negative lens are needed. Because of the negative lens we have negative aberrations and, therefore, aberration canceling becomes more favorable than when combining two positive lenses. Then it is possible, with the combination of two achromatic doublets, one positive and one negative, to correct for all the primary aberrations.

We select two thin achromatic doublets with the same glass choice and opposite optical power, which results in $\sigma_{IVA} = -\sigma_{IVB}$, and Petzval field curvature is corrected. The aperture stop is set at the front positive doublet, as shown in Figure 8.18, where lens thickness has been included. There is no distortion aberration from the front doublet because the aperture stop coincides with the doublet, and astigmatism aberration is fixed. The front doublet, if not cemented, can be corrected for spherical aberration and coma aberration.

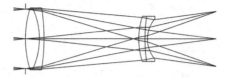

Figure 8.18 Telephoto lens with BK7 and SF5 glasses. $f' = 100$ mm, $F/4$, FOV = $\pm 6.2°$, TTL/$F = 0.8$.

However, with the glass choice of BK5 and SF5, the front doublet can be cemented and still be aplanatic. The rear doublet can also be cemented and aplanatic. The astigmatism of the rear doublet with remote stop is given by, $S_{III}^* = S_{III} + 2 \cdot \bar{S} S_{II} + \bar{S}^2 S_I$, but since the doublet is aplanatic we have that, $S_{III}^* = S_{III} = \mathcal{K}^2 \phi_B \sigma_{IIIB} = \mathcal{K}^2 \phi_B$. It follows then that, because both doublets have opposite optical power, $\phi_A = -\phi_B$, the total astigmatism is zero. In cases where the rear doublet cannot be corrected for coma aberration, residual coma in the front doublet can be used to make the combination aplanatic. Some coma in the rear doublet can be used to correct astigmatism aberration. Thus, we have a telephoto lens made out of two achromatic and aplanatic, or nearly aplanatic, doublets that is anastigmatic. This, however, suffers from a small amount of pincushion distortion.

In order to correct for distortion aberration, it is necessary to increase the degrees of design freedom by not using cemented doublets. However, and again by the proper choice of glass, BK7 and F6, it is possible to keep the front doublet as cemented. The front doublet does not introduce distortion; the contribution is from the rear doublet. Since distortion aberration is related to pupil coma, $\bar{W}_{131} = W_{311} + \frac{1}{2} \mathcal{K} \cdot \Delta \{ \bar{u}^2 \}$, we can understand that bending the rear doublet provides a degree of freedom for correcting distortion, as pupil coma is a linear function of lens bending. The change of coma in the image is compensated by bending (or by the glass choice) the front doublet. The airspace between doublets is used to restore the correction for astigmatism. Astigmatism of the second doublet depends on the stop position, since this doublet is no longer aplanatic.

Figure 8.19 shows a telephoto lens corrected for all primary aberrations including distortion, and that was further optimized by minimizing the RMS spot size across the field of view, as shown in Figure 8.20. There are several glass choices of crown-in-front, or flint-in-front, for the doublets. The telephoto lens in Figure 8.19 has crown in front in both doublets, providing the best performance among other doublet combinations possible. Table 8.2 provides the constructional data of the telephoto lens. Note that there is no lens spacing in the rear doublet, and the lens elements have center contact. This is not good practice,

Table 8.2 *Data of the telephoto lens (mm).* f′ = *100 mm, F/4, FOV = ±6.2°,* *TTL/F = 0.8. Stop at first surface*

Surface	Radius	Thickness	Glass
1	30.8260	6	BK7
2	−37.0680	2	F6
3	−1,085.8357	47.8444	
4	−16.0627	1.0648	BK7
5	143.4857	0	
6	35.6724	2.6621	F6
7	3,254.9727	20.8462	

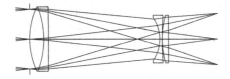

Figure 8.19 Telephoto lens with BK7 and F6 glasses. $f′$ = 100 mm, F/4, FOV = ±6.2°, TTL/F = 0.8.

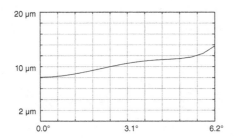

Figure 8.20 RMS spot size across the field of view of the optimized telephoto lens.

because the lenses can get damaged. The telephoto then would be re-optimized with a 0.2 mm lens spacing in the rear doublet. The telephoto ratio is TTL/$f′$ = 0.8. The field of view (FOV) is ±6.2° and is small in comparison to the previous lenses discussed, but the optical speed is F/4. There is room to trade-off field of view vs. optical speed. However, light vignetting by the rear doublet can be of concern. It is desirable to keep the stop aperture at the front doublet to minimize its size and, thus, keep at a minimum the diameter of the telephoto. For the optimized telephoto lens, the final doublet focal lengths are 70 mm and −56 mm for a combined focal length of $f′$ = 100 mm.

8.9 Reverse Telephoto Lens

As its name implies, a reverse telephoto lens has a negative lens followed by a positive lens. This combination results in an increased distance between the last lens surface, known as the back focal length (BFL), and the image plane, as shown in Figure 8.21. The aberration correction rationale to design a reverse telephoto lens out of two achromatic doublets is the same as for the telephoto lens. Both doublets are chosen with opposite optical powers to cancel the Petzval sum. The rear doublet works at nearly negative unit magnification, $m = -1$. In addition, both doublets are made aplanatic and cemented by lens bending and by the choice of glass. Then astigmatism aberration also cancels. Some barrel distortion of about -1.5% remains uncorrected in such a simple combination that is not symmetrical about the stop aperture. The stop is set at the positive doublet to help control aberration and minimize lens diameter. The RMS spot size across the field of view is shown in Figure 8.22.

The reverse telephoto lens in Figure 8.21 has a focal length of $f' = 100$ mm, a total track of 324 mm, and back focal length of 192 mm, works at $F/4$, and has a field of view of FOV = $\pm12°$. The focal lengths of the doublets are -100 mm and 100 mm. The glass choice is BK7 and SF5 from Schott. The feature of the reverse telephoto lens is the large back focal length at the expense of a large total track length.

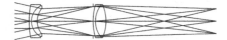

Figure 8.21 Reverse telephoto lens with BK7 and SF5 glasses. $f' = 100$ mm, BFL = 200 mm, TTL = 324 mm, FOV = $\pm12°$, $F/4$.

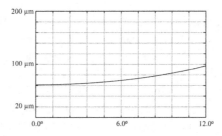

Figure 8.22 Reverse telephoto lens rms spot size across the field of view. $f' = 100$ mm, BFL = 200 mm, TTL = 324 mm, FOV = $\pm12°$, $F/4$.

Further Reading

Dallmeyer, J. H. Photographic Lens, U.S. Patent 79,323 (1868).

Gao, Weichuan, Sasián, José. "Air lens vs aspheric surface: a lens design case study," Proceedings of SPIE 10590, International Optical Design Conference 2017, 105900B (2017); doi: 10.1117/12.2287888.

Kingslake, R. "Telephoto vs. ordinary lenses," *Journal of the SMPTE*, 75 (1966), 1165–68.

Kingslake, R. "The reversed telephoto objective," *Journal of the SMPTE*, 75 (1966), 203–7.

Kingslake, R. *A History of the Photographic lens* (San Diego, CA: Academic Press, 1989).

Lister, J. "On some properties of achromatic object-glasses applicable to the improvement of the microscope," *Philosophical Transactions of the Royal Society*, Part I (1830), 187–200.

Lummer, Otto. *Contributions to Photographic Optics* (London: MacMillan and Co., 1900).

Merte, W. Telephoto Objective, U.S. Patent 1,467,804 (1923).

Richter, R. Photographic Teleobjective, U.S. Patent 2,239,538 (1941).

Rudolph, P. Aplanatic Lens, U.S. Patent 444,714, Photographic Objective (1891).

Sasián, José. "Joseph Petzval lens design approach," Proceedings of SPIE 10590, International Optical Design Conference 2017, 1059017 (2017); doi: 10.1117/12.2285108.

Schroder, H. L. H. Aplanatic Lens, U.S. Patent 404,506 (1889).

9

Image Evaluation

The quality of a lens system for a given application depends on several factors. Among them are the choice of lens form that best suits the application, the image quality provided by the lens, and how well the image quality can be maintained under actual lens fabrication and assembly errors, environmental conditions, and actual lens use. A starting point for a lens design is the lens specifications, which includes first-order requirements, packaging constraints, optical power-efficiency, and image quality requirements. Ultimately, lens cost is often a major design driver.

It is of critical importance that the application for the lens is well understood throughout, as much as possible, as the specifications for that lens follow from that understanding. In some cases, the specification list is wrong or incomplete, and this obviously impacts the lens design; it is often up to the engineer to discuss the specifications with the customer to correct any deficiency. The specifications may define the expected image quality in terms of image evaluation metrics that depend on the system application.

Once a lens form is chosen and the aberrations have been corrected, minimized, or balanced, it is necessary to evaluate how good or bad the image cast by that lens could be. This chapter discusses a variety of concepts and tools in image evaluation that are essential in lens design.

There are several topics that are related and that are of interest to understand and develop skill in their use as they relate to lens design. These topics can be treated from a geometrical optics point of view or from a physical optics point of view and are: image formation theory, image aberration theory, aberration evaluation, and image evaluation.

This chapter first discusses the image evaluation of a point object, and second the image evaluation of an extended object.

9.1 Image Evaluation of a Point Object

The concept of a point object, or light source, is non-physical, as it does not have an area. However, it is a useful concept, in that it defines where light rays or waves depart from. Consequently, and geometrically, the ideal image of a point object is a point image. The basic idea of image evaluation is that an image is composed of the images of a plurality of object points and, by assessing the individual point images, insight is obtained about the overall image quality of an extended object.

Observation of the images of point objects produced by aberration free lens systems with circular entrance pupils shows that the images are not pointy, but have a light distribution known as the Airy pattern, shown in Figure 9.1.

This pattern, $A(\rho)$, is described mathematically by the function,

$$A(\rho) = \left(\frac{2J_1(\pi\rho)}{(\pi\rho)}\right)^2, \tag{9.1}$$

where $J_1(\rho)$ is a first-order Bessel function of the first kind. The radial locations of the first three zeros in the Airy pattern take place at approximately $\rho = 1.22, \rho = 2.23$, and $\rho = 3.24$. The radius of the first zero as it relates to the wavelength of light and to the F-number is given by $\rho = 1.22\lambda F/\#$. Thus, taking the F-number of the eye to be $F/2.8$ and for a wavelength of $\lambda = 555$ nm, the radius of the first dark ring in the Airy pattern is $\rho = 1.89$ μm. Figure 9.2 shows a cross-section of the Airy pattern, and Figure 9.3 shows the encircled energy. About 83% of the energy is encircled within the first dark ring, and 91% within the second dark ring.

The Airy pattern occurs because of the wave nature of light and the phenomenon of diffraction. For many applications, the Airy pattern is considered as the ideal image of a point object that an aberration-free system can produce. In practice, one requirement for a system to produce an image that is substantially the Airy pattern is that the wavefront deformation across the exit pupil must be small in comparison to the wavelength of light.

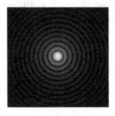

Figure 9.1 Enhanced Airy pattern produced by an aberration-free optical system.

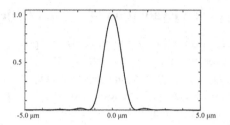

Figure 9.2 Normalized cross-section of the Airy pattern for $\lambda = 1$ μm and $F/\# = 1$.

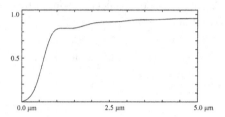

Figure 9.3 Normalized encircled energy in the Airy pattern as a function of the radial distance for $\lambda = 1$ μm and $F/\# = 1$.

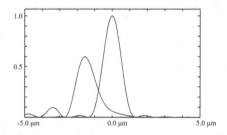

Figure 9.4 Cross-section of the diffraction pattern for the case of 1λ of coma aberration, with a peak of 0.6 (left), and no aberration with a peak of 1.0 (right) for $\lambda = 1$ μm and $F/\# = 1$.

One major effect of aberration in the Airy pattern is to decrease the peak irradiance, as shown for the case of 1λ of coma aberration in Figure 9.4.

The ratio of the peak irradiance of the diffraction pattern produced by an aberrated beam to the peak irradiance of an un-aberrated beam was first estimated by Lord Rayleigh, but is known as the Strehl ratio. This ratio, R, was shown by A. Marechal to be related to the variance of the wavefront deformation, σ_W^2, and can be approximated by the R. Shack relationship,

$$R = \exp\left(-\left(\frac{2\pi}{\lambda}\right)^2 \sigma_W^2\right), \tag{9.2}$$

Table 9.1 *Primary aberration that produces a ratio of $R = 0.8$ at the ideal image plane, $\sigma_W \cong 0.07\lambda$*

$W_{020} = 0.247\lambda$	$W_{040} = 0.239\lambda$	$W_{131} = 0.604\lambda$	$W_{222} = 0.319\lambda$

whenever the ratio satisfies $R \geq 0.1$. An optical system where the ratio is larger than 0.8 is considered to be very good, and is called diffraction limited. For this the standard deviation, σ_W, of the wavefront must be less than 0.07λ. Table 9.1 provides the amount of primary aberration that would produce a ratio of $R = 0.8$.

For the primary aberrations, the variance of the wavefront as a function of the field of view is given by,

$$\sigma_W^2 = \begin{pmatrix} \dfrac{1}{12}\left(W_{020}+W_{040}+\left(W_{220}+\dfrac{1}{2}W_{222}\right)\vec{H}\cdot\vec{H}\right)^2 + \dfrac{1}{180}W_{040}^2 \\ +\dfrac{1}{24}W_{222}^2\left(\vec{H}\cdot\vec{H}\right)^2 + \dfrac{1}{4}\left(\dfrac{2}{3}W_{131}\left|\vec{H}\right|+W_{311}\left(\vec{H}\cdot\vec{H}\right)\left|\vec{H}\right|\right)^2 + \dfrac{1}{72}W_{131}^2\vec{H}\cdot\vec{H} \end{pmatrix}.$$

(9.3)

The difference, $E = 1 - R$, can be considered the amount of energy that is taken out of the Airy disk and redistributed among the diffraction rings.

The Airy pattern is considered an ideal image and serves as a benchmark for comparison purposes. A useful analogy to keep in mind is that the diffraction pattern produced by a lens system, or light spot, is like the point of a pencil; the sharper it is, the finer the lines it can draw. Therefore, for actual systems it is important to evaluate the sharpness of the light spots they can produce. There are a variety of tools available to a lens designer to estimate how sharp the image of a point can be. For example, ray fans, wave fans, spot diagrams, RMS wavefront deformation plots, and encircled energy plots. These metrics do depend on the wavelength(s) the lens system is intended to work and, therefore, it is important to include in any analysis the spectral bandwidth of the system by weighting the contributions of a number of wavelengths.

9.1.1 Resolving Power and Depth of Focus

Two important attributes of an optical system are its resolving power and its depth of focus. Resolving power (RP) refers to the ability to separate two objects that are close to each other. Depth of focus (DOF) refers to the ability to maintain spot size as the observation plane is moved along the optical axis.

Figure 9.5 Images of two point objects. Left, not resolved; center, resolved according to Rayleigh's criterion; right, resolved.

For a system that is diffraction limited, these can be expressed in the terms of the wavelength, λ, and numerical aperture, NA, of the system as,

$$RP = k_1 \frac{\lambda}{NA}, \tag{9.4}$$

and

$$DOF = \pm k_2 \frac{n\lambda}{NA^2}, \tag{9.5}$$

where k_1 and k_2 are parameters that depend on the application, and n is the index of refraction in image space. As shown in Figure 9.5, Rayleigh's resolving power criterion refers to the ability to barely resolve the images of two point objects of equal intensity, and sets $k_1 = 0.61$. This criterion has the peak of the image of a first point falling on the first zero of the image of a second point. A defocus amount of $W_{020} = \lambda/4$ is considered tolerable and sets $k_2 = 0.5$. For a diffraction limited system working at $\lambda = 0.5$ μm, a rule of thumb is that the DOF is equal to the square of the $F/\#$ in micrometers. Thus, an $F/10$ system would have a DOF of ± 100 μm. For systems that are not diffraction limited, both the resolving power and the depth of focus will depend on the application.

9.1.2 Spot Diagram

Spot diagrams are computed by plotting the intersection point of rays with the observation surface for a given field point, and, thus, they are a purely geometrical optics construct. This is shown in Figure 9.6 (left) for the case of five waves of coma aberration. Spot diagrams accompanied by a scale bar, or compared to the Airy disk diameter, provide a first estimate of the image size of a point. In addition, they can provide information about the symmetry of the aberration in a light beam. However, they do not convey how the energy is distributed spatially, unless the number of rays per bin is counted, and plotted as shown in Figure 9.6 (center). Spot diagrams can also be computed

Figure 9.6 Spot diagrams for five waves of coma aberration. Left, standard spot; center, binning and counting rays; right, diffraction calculated.

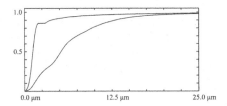

Figure 9.7 Encircled energy vs. radial distance for a diffraction limited system, left, and for the on-axis beam of the Petzval portrait objective, right.

using light diffraction calculations, and then a more accurate representation of the light distribution is obtained, as shown in Figure 9.6 (right). The rms spot size across the field of view is also a first estimate of the image quality of a system.

9.1.3 Encircled Energy

A plot of the encircled energy as a function of spot radius is an important tool to evaluate the quality of the image of a point object. Direct comparison with the encircled energy of a diffraction-limited beam provides objective information. Figure 9.7 shows the encircled energy for the on-axis beam of the Petzval portrait objective calculated for the F, d, and C wavelengths, and when the objective is scaled to a focal length of 100 mm. About 85% of the energy is encircled within a radius of 11.0 μm, while in the diffraction-limited beam 85% is encircled in a radius of about 3.0 μm.

The fact that the spot of light produced by the Petzval objective is much larger, about four-times, than the diffraction limited spot, does not mean that it is a poor lens. Considering that photographic film as a rough rule of thumb has a grain size of 25.0 μm, then the on-axis spot of the Petzval objective is not a limiting factor for image quality, Thus, the required image quality depends on application.

Some applications may require plots of ensquared energy, or energy enclosed in the meridional or sagittal directions, as a function of the field of

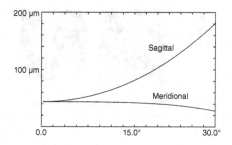

Figure 9.8 Plots of distance in the meridional or sagittal directions to enclose 80% of energy (geometrical) as a function of the field of view for the Wollaston landscape lens, $f' = 100$ mm.

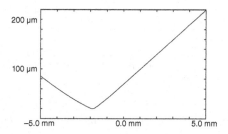

Figure 9.9 Radius that encircles 80% of the energy (geometrical) vs. axial position for the on-axis beam of the Wollaston landscape lens.

view, as shown in Figure 9.8 for the Wollaston landscape lens. Figure 9.9 also shows, for the Wollaston lens, the radius which encircles 80% of the energy (geometrical) for the on-axis beam as a function of the axial position. This provides information about the depth of focus.

9.2 Image Evaluation of an Extended Object

Application of linear shift invariant theory to model imaging by a lens system resulted in an insightful and powerful way to evaluate images of extended objects. In a linear shift invariant system (LSIS), the image, $i(x, y)$, is the convolution of the object, $o(x, y)$, with the system point spread function, $h(x, y)$, which is also called the impulse function. Mathematically this is written as,

$$i(x, y) = o(x, y) * * h(x, y). \tag{9.6}$$

An object can be analyzed and described as a superposition of a plurality of spatial frequencies, $\exp\{-i2\pi(\xi x + \eta y)\}$, and likewise the image can

Figure 9.10 Spatial frequency pattern. Left, with full contrast; center and right, with reduced contrast; right, with contrast reversal.

also be analyzed and described by a plurality of spatial frequencies, $\exp\{-i2\pi(\xi x + \eta y)\}$. For example, an object represented by a square wave can be described by a superposition of cosine waves with different spatial frequencies, amplitudes, and phase shifts.

By taking the Fourier transform of the image, $i(x, y) = o(x, y) ** h(x, y)$, we can write,

$$I(\xi, \eta) = O(\xi, \eta) \times H(\xi, \eta), \tag{9.7}$$

where $O(\xi, \eta)$ represents the spatial frequency spectrum of the object, $I(\xi, \eta)$ is the spatial frequency spectrum of the image, and $H(\xi, \eta)$ is the optical transfer function (OTF) of the system.

In general, the optical transfer function is complex, and can be written as,

$$H(\xi, \eta) = MTF(\xi, \eta) \times \exp\left(-i\varphi(\xi, \eta)\right), \tag{9.8}$$

where $MTF(\xi, \eta)$ is its modulus, and $\varphi(\xi, \eta)$ is the phase. The modulus, $MTF(\xi, \eta)$, is called the modulation transfer function, and provides the contrast with which a given spatial frequency is imaged. For a periodic function, the contrast is given by,

$$C(\xi, \eta) = MTF(\xi, \eta) = \frac{\text{MAX} - \text{MIN}}{\text{MAX} + \text{MIN}}, \tag{9.9}$$

where MAX and MIN are the maximum and the minimum irradiances. Figure 9.10 (left) shows one spatial frequency with which an object can be represented. As it is imaged by an optical system, the contrast can be reduced, as shown in Figure 9.10 (center and right). In addition, there can be a phase shift, i.e. a lateral displacement of the frequency pattern, that, when it is 180°, produces contrast reversal, as shown in Figure 9.10 (right). In contrast reversal, the peak irradiance displaces by half a period so that peaks and valleys swap location. Axially symmetric systems that are linear shift invariant can only produce phase shifts of 0° or 180°.

Under incoherent object illumination and for an aberration free system that has a circular entrance pupil, the point spread function, $h(x, y)$, is the Airy pattern. Its Fourier transform represents an ideal optical transfer function and, since for this case there is no phase shift, $\varphi(\xi, \eta) = 0$, we have $H(\xi, \eta) = MTF(\xi, \eta)$ and,

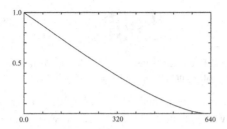

Figure 9.11 Modulation transfer function of an aberration-free system with circular entrance pupil working at *F*/2.8 and at λ = 555 nm. The vertical axis is the contrast, and the horizontal axis is the spatial frequency, $f = \sqrt{\xi^2 + \eta^2}$ in lp/mm.

$$MTF(\xi, \eta) = \frac{2}{\pi}(\psi - \cos(\psi)\sin(\psi)), \qquad (9.10)$$

where

$$\psi = ar\cos\left(\frac{\lambda\sqrt{\xi^2 + \eta^2}}{2NA}\right) = ar\cos\left(\frac{f}{f_c}\right). \qquad (9.11)$$

The MTF is a three-dimensional function as a function of the spatial frequencies, ξ and η. However, for an aberration-free system that is axially symmetric, the MTF is also axially symmetric, and a cross-section is sufficient to describe it, as shown in Figure 9.11. There is no modulation after the cut-off frequency, $f_c = 2NA/\lambda$; this is, the system cannot image spatial frequencies beyond this frequency, f_c. For a system in air working at *F*/2.8 and at λ = 555 nm, the cut-off frequency is $f_c \cong 643.5$ line-pairs per millimeter (lp/mm).

The MTF shows that only the zero spatial frequency is imaged with full contrast, and that the contrast progressively decreases as the spatial frequency increases. The MTF provides useful information about how an extended object is imaged. It tells what frequencies can be imaged and with what contrast. Clearly, this provides a different perspective about how to evaluate an image.

For off-axis field points, the MTF is no longer axially symmetric. Different cross-sections of the three-dimensional MTF provide the contrast of spatial frequencies that have different orientations with respect to the optical axis. MTF cross-sections in the meridional and sagittal directions are often plotted by lens design programs.

One can understand the OTF by imaging with a lens an object having a single spatial frequency, such as in Figure 9.10 (left), and measuring the contrast and phase shift in the image. By rotating the object and measuring contrast and phase shift, the OTF for that frequency can be determined. The process can be repeated for many more spatial frequencies to determine the OTF of the lens more completely.

9.2.1 MTF Curves

Optical aberrations degrade the MTF, as shown in Figure 9.12, for example, for the case of defocus: $W_{020} = \lambda/4$, $W_{020} = \lambda/2$, and $W_{020} = \lambda$. For 1λ of defocus the phase (not shown) becomes negative after the modulation reaches zero and contrast reversal takes place. Defocus has a severe and adverse effect on the MTF, and about $W_{020} = \lambda/4$ can be tolerated in a diffraction-limited system.

The MTF is also plotted for the Wollaston landscape lens in Figure 9.13. The diffraction-limited MTF is plotted for reference, then the MTF for the on-axis field point, and finally two cross-sections in the meridional (tangential) and sagittal direction are plotted for the $30°$ field of view. One must keep in mind that the MTF is a three-dimensional function, and that plotting a few cross-sections of it may not be sufficient. However, it is standard practice in lens design to plot the S and T cross-sections. The Wollaston landscape lens is far from performing at the diffraction limit, but is still a useful imaging lens.

A healthy human eye is said to have a resolving power of about 1 arc-minute. But, in consideration of several degrading factors, an average

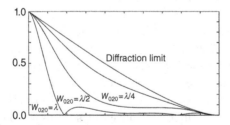

Figure 9.12 Diffraction-limited MTF and in the presence of defocus W_{020}.

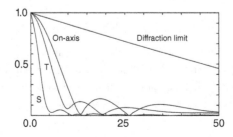

Figure 9.13 MTF for the Wollaston landscape lens.

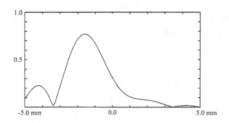

Figure 9.14 MTF for a spatial frequency of 10 lp/mm vs. axial position for the Wollaston landscape lens, $f' = 100$ mm. The zero position corresponds to the ideal image plane location, as calculated with first-order ray tracing.

resolving power can be set to 3 arc-minutes. Then, when looking at a photograph at a distance of 250 mm, the eye can resolve two spots that are separated by about 0.25 mm. This corresponds to an object spatial frequency of $f = 4$ lp/mm.

For 35 mm film (24 mm × 36 mm) and assuming that the printed photograph would be 8-times larger (192 mm × 288 mm), and viewed at a distance of 250 mm, the requirement for spatial frequency would be 4 lp/mm in the printed photograph and 32 lp/mm in the 35 mm film. Thus, a rule of thumb to specify a lens for 35 mm film is a contrast larger than 50% at 32 lp/mm. For high quality lenses the rule of thumb stiffens and becomes a contrast of 50% at 30 lp/mm and a contrast of 30% at 50 lp/mm. This is sometimes referred to as the 30/50 rule. However, the specification for MTF is highly dependent on application.

For the current CCD and CMOS image sensors with pixel sizes in the order of a few micrometers, the spatial frequency requirements are even higher. For example, to take advantage of a pixel size of 5 μm, the lens must be able to produce spatial frequencies in the order of 100 lp/mm.

The MTF for a given spatial frequency and for a given field can be plotted as a function of the axial position of the image plane. This is shown in Figure 9.14 for the Wollaston landscape lens for a spatial frequency of 10 lp/mm. Plots like this provide information about the best imaging plane and the depth of focus.

9.2.2 Image Simulation

The irradiance of an image can be calculated, and an image simulation can be produced in great detail under a number of variables, such as the illumination, the system aberrations, and the nature of the object to be imaged. An equation that describes the process of physical image formation was developed by H. H. Hopkins and is,

$$I(x,y) = \left(\frac{1}{f\lambda}\right)^2 \iint_\infty \sigma(x_0, y_0)|s(x,y)t(x,y) * *psf(x,y)|^2 dx_0 dy_0, \qquad (9.12)$$

Figure 9.15 Left, bar target as an object. Right, image simulation using a Petzval lens. Note the decrease in relative illumination toward the image corners, and the pincushion distortion.

where $I(x, y)$ represents the irradiance of the image, $\sigma(x_0, y_0)$ represents the illumination, $s(x, y)$ is the optical field, $t(x, y)$ represents the object transmittance, and $psf(x, y)$ is the point spread function of the system. For incoherent illumination, the Hopkins equation simplifies to a convolution operation that can be carried with the fast Fourier transform. For partial coherent illumination, the equation is no longer a convolution, and the equation would take more time to be computed. Image simulation with the Hopkins equation can take significant time. However, it is used to calculate fine details in the image that an optical system could form.

As an example of image simulation, Figure 9.15 (left) shows a bar target as an object. Figure 9.15 (right) shows the simulated image that a Petzval portrait lens would produce. There is a decrease in relative illumination and in resolving power toward the corners of the image. Image simulations are helpful for a variety of purposes, such as checking image orientation, the overall appearance of the image, illumination, and resolving power.

Further Reading

Airy, G. B. "On the diffraction of an object-glass with circular aperture," *Transactions of the Cambridge Philosophical Society*, 5 (1835), 283–91.

André, Maréchal. "Etude des effets combinés de la diffraction et des aberrations géométriques sur l'image d'un point lumineux," *Review Optics*, 2 (1947), 257–77.

Barrett, H., Myers, K. *Foundations of Image Science* (New York: John Wiley & Sons, 2004).

Gaskill, J. D. *Linear Systems, Fourier Transforms, and Optics* (New York: John Wiley & Sons, 1978).

Hopkins, H. H. "On the diffraction theory of optical images," *Proceedings of the Royal Society A*, 217 (1952), 408–31.

Shack, R. V. Interaction of an Optical System with the Incoming Wavefront in the Presence of Atmospheric Turbulence, Optical Sciences Center, Technical Report 19, The University of Arizona (1967).

Shannon, R. R. *The Art and Science of Optical Design* (Cambridge, UK: Cambridge University Press, 1997).

Strutt, John William, Lord Rayleigh. "Investigations in optics, with special reference to the spectroscope," *Philosophical Magazine*, 5(8) (1879), 403–11.

10

Lens Tolerancing

A lens manufacturer requires tolerances in the dimensions of a lens to be able to provide a cost estimate and be able to manufacture the lens. Further, for the lens to meet the lens specifications after it is built, it is necessary that the actual lens dimensions do not depart from the nominal design ones by some amounts known as fabrication and assembly tolerances. Thus, the task of the lens designer is not only to provide a lens design that meets image quality requirements, but to also provide tolerances, so that the as-built lens actually meets the specifications and satisfies the needs of the application. Critical goals of the lens tolerancing process are to provide tolerances to each of the constructional parameters of the lens, and to find out the statistics of the as-built lens so that the fabrication yield, and final cost, can be estimated. This chapter provides a primer into the lens tolerancing process. Commercial lens design software allows for the lens tolerancing analyses discussed below.

10.1 Lens Dimensions and Tolerances

A lens designer needs to develop an understanding of physical dimensions and their measurement so that realistic tolerances can be assigned. He or she needs to have insight into linear and angular dimensions, such as how big a micrometer is, or one-arc second is. In lens fabrication, both of these magnitudes often separate what is very difficult to make from what is reasonable to make. One must find out how a given lens dimension will be achieved and measured in the optics shop. If it cannot be measured, it probably cannot be made to specification.

Twenty-five micrometers (25 μm) is a useful reference. The minimum measurement division of many instruments and machining tools is 0.001″, or about 25 μm. Asking for an optical element to be made with a tolerance of

Table 10.1 *Tolerance guidelines for glass lenses, 10 mm to 100 mm in diameter*

Lens parameter	Low precision	Precise	High precision	Requires special process
Diameter (mm)	+0.0 −0.25	+0.0 −0.1	+0.0 −0.025	+0.0 −0.005
Central thickness (mm)	±0.12	±0.05	±0.012	±0.002
Edge thickness difference (mm)	0.12	0.012	0.006	0.003
Surface radius (rings)	5% (10)	1% (3–5)	0.1% (1)	0.01% (0.25)
Wavefront error from surface figure	0.5λ RMS	0.07λ RMS	0.04λ RMS	0.02λ RMS

25 μm is considered doable. Asking for that element to be made to 50 μm or more is considered easy. However, asking for an optical part to be made to 12.5 μm starts to become difficult, to 2.5 μm becomes very difficult, and to 0.25 μm extremely difficult. Similarly, by dividing 25 μm over a lens diameter of 25 mm, we get an angular tolerance of about 3.3 arc-minutes, which is doable. One order of magnitude up or down makes the angular tolerance easy or difficult to achieve.

Different optics shops can make a given lens dimension, such as lens diameter, lens thickness, surface radius, or wedge between the lens surfaces, with a tight tolerance for a given cost, or cannot achieve a given tolerance. The lens designer needs to have effective communication with the lens manufacturers, to find out how well they can achieve lens tolerances, and their associated cost. Lens manufacturers provide guidelines for the different lens tolerances they can achieve under some assumptions. Generally, the tighter the tolerances, the costlier the lens becomes. What a tight tolerance is also depends on the manufacturing process. For example, state-of-the-art, mass produced injection molded lenses for mobile phones routinely achieve micrometer level tolerances. Table 10.1 provides some guidelines for the level of tolerances for lenses with spherical surfaces in the order of 10–100 mm in diameter, made out of glass, and which are not mass produced.

The lens diameter refers to the actual lens diameter, in comparison to the clear aperture of the lens that performs the optical function of refracting or reflecting light rays. A common surface polishing problem is to have the very edge of the surface turned down. To overcome this figuring problem, there is a tendency to specify a lens diameter larger, say 10–20% larger, than the clear aperture. However, usually packaging requirements and lens cost win and the diameter of the lens is minimized to only allow for enough clearance to properly mount the lens. It is imperative that a bevel, or protective chamfer, is specified to avoid the lens edge easily chipping.

The central lens thickness is measured from surface center to surface center, i.e. along the optical axis. Measuring central thickness requires finding the central portion of the lens, and this contributes to making a precise measurement difficult.

Edge thickness difference, or lens wedge, is measured by supporting the lens in a kinematical mount so that its position is well-defined, and rotating the lens while a micrometer measures the position of the lens edge as the lens rotates. This produces the micrometer reading to oscillate between a minimum and a maximum value, which is the edge thickness difference, called the total indicator runoff. This difference, divided by the lens diameter, gives the lens wedge.

Measuring the radius of curvature of a surface requires an optical bench. Alternatively, optics shops have a collection of test plates with radii of curvature measured with accuracy in an optical bench. Then the lens designer, in a final lens optimization run, fits the radii of curvature of the surfaces of the lens to the radii of curvature of the optics shop test plates. The optics shop tests for radii of curvature errors by observing the Newton rings formed by the test plate and a given lens surface. In this method, the surface radius of curvature is given a tolerance in Newton rings at a given wavelength of light. One Newton ring represents $1/2\lambda$ of sag difference at the edge of the lens between the test plate and the lens surface.

Surface figure, or irregularity, refers to the departure of a surface from the spherical shape, or from the nominal designed aspheric shape. There are many types of figure error, such as surface cylindrical deformation, which would introduce astigmatism aberration, an axially symmetrical error, which would introduce spherical aberration, such as turned down edge, periodic surface errors, which could diffract light and introduce image artifacts, asymmetric surface errors, and others. These figure errors depend on the lens manufacturing method. For example, single point diamond turning produces periodic high spatial frequency figure errors.

A change in the glass index of refraction of a lens element will change the first-order properties of a lens system and will introduce wavefront changes. A change in the glass v-number of a lens element will change the chromatic correction. To minimize errors, the index of refraction of the glass to be used in the lens manufacture is measured, and the lens is re-optimized to reflect the actual index of refraction. For optical systems with glass elements larger than about 80 mm in diameter, and that are diffraction limited, index of refraction homogeneity within the glass is also a concern.

Figure 10.1 Parameter error distributions. From left to right, uniform, end limited, truncated normal, skewed, parabolic.

Each of the constructional parameters of a lens can have a given error distribution. For example, the error in central lens thickness may be biased to the thicker side to allow room for regrinding a lens in case the surface becomes scratched. Some parameter error distributions are uniform, end-limited, truncated normal, shifted-skewed, and parabolic, as shown in Figure 10.1.

10.2 Worst Case

It is perhaps tempting to try to determine the worst case performance of a lens that will be manufactured under a variety of fabrication errors. Determining the worst case estimate is not practical, because it would require us to compute the effects of all combinations of errors, and this can take an excessive amount of time, even for simple systems.

Alternatively, if there are, say, n causes of errors, a worst case can be set by adding all the effects of the errors in the same direction. However, this approach is pessimistic.

Therefore, the approach that is taken in practice for tolerancing is statistical in nature. Consequently, one goal in tolerancing a lens is to estimate the statistics of the as-built lens.

10.3 Sensitivity Analysis

For tolerancing a lens it is necessary to define a criterion of performance such as, for example, the error function used to optimize the lens. It is important to properly reflect relevant aspects of the lens in the tolerancing criterion. An insufficient criterion may lead to a faulty tolerancing analysis.

A sensitivity analysis uses a list of all the constructional parameters that can have actual fabrication errors, such as lens thickness, lens spacing, surface curvature, and index of refraction. Then, tolerances are assigned and used to vary the constructional parameters of a lens, one at each time, and determine how much the tolerancing criterion has changed. This is done for each of the

Table 10.2 *Sensitivity analysis*

Surface	Item	Nominal value	Tolerance	Criterion change
1	Radius	50 mm	0.01 mm	0.3
2	Thickness	8 mm	0.1 mm	0.005
3	Index	1.51	0.001	0.001

Table 10.3 *Inverse sensitivity analysis*

Surface	Item	Nominal value	Tolerance	Criterion change
1	Radius	50 mm	0.003 mm	0.01
2	Thickness	8 mm	0.2 mm	0.01
3	Index	1.51	0.01	0.01

constructional parameters, and the changes in the criterion are ranked to determine which parameters produce the larger changes in the criterion. Table 10.2 provides an example of the data produced in a sensitivity analysis.

A sensitivity analysis produces two useful pieces of information: the lens parameters that worst offend the performance of the lens, and the criterion changes which can be used to estimate the statistics of the as-built lens.

10.4 Inverse Sensitivity Analysis

In an inverse sensitivity analysis, tolerances are determined that would produce a given change in the tolerancing criterion. Table 10.3 provide an example of the data produced in an inverse sensitivity analysis. Such analyses provide information about the levels of tolerance needed for a given performance, and indicate which parameters require loose or tight tolerances.

10.5 Compensators

In order to relax tolerances and reduce manufacturing cost, some compensators such as an air-space, or a lens decenter, can be used to improve a lens system after the lens elements have been made. For example, the back focal length is used to best focus the image, and an airspace can be used to restore the focal length or to correct for residual spherical aberration. However, for mass produced lenses, it is desirable to not specify compensators, as their

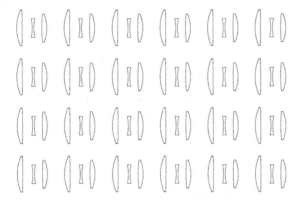

Figure 10.2 Twenty-four Cooke triplet lenses

implementation requires testing and time to fix the problem. Best focusing of the lens by moving the lens assembly, or the image sensor, is most often specified as a compensator.

10.6 Tolerancing Criterion Statistics

Often lenses are manufactured in bulk, and the quality of each lens differs among the lenses because the manufacturing errors are not the same for all the lenses. Or, even, a single lens system where the lens is disassembled and reassembled, can result in a different lens because the lens element positions and air spaces vary. Figure 10.2 shows twenty-four Cooke triplet lenses. If the performance of these lenses were to be measured, one would find variation in the focal length and in the image quality.

Theory shows that, when the manufacturing errors are very small, and for a given tolerancing criterion such as the RMS spot size, or RMS wavefront error, the histogram for a large number of lenses approaches a normal probability distribution, as shown in Figure 10.3 (left). However, in practice, as the errors are not very small, the histogram is skewed, as shown in Figure 10.3 (right).

A reason for why, under very small errors, the histogram tends to be approached by a normal distribution is the central limit theorem. This theorem states that, for a set of independent and random variables having a mean and a variance, the probability density function of the sum of the variables approaches a normal distribution as the number of variables increases. A reason for why the histogram becomes skewed when the errors become larger is that, as the lens has been optimized, most combinations of changes

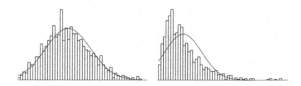

Figure 10.3 Left, histogram of RMS spot size for 1,000 Cooke triplets under very small fabrication errors. Right, histogram of the Cooke triplets under large fabrication errors. A best fit normal distribution has been overlaid with the histograms.

will tend to degrade the performance, and very few, or none, will tend to improve it.

For simplicity, a first estimate for the probability density function, $p(S)$, of a tolerancing criterion, S, is a normal distribution defined by,

$$p(S) = \frac{1}{\sigma_S\sqrt{2\pi}} \exp\left\{\frac{-(S - \langle S\rangle)^2}{2\sigma_S^2}\right\}, \tag{10.1}$$

where $\langle S\rangle$ is the mean, and σ_S^2 is the variance. The mean can be estimated by,

$$\langle S\rangle = S_0 + \sum_{i=1}^{j} \langle \Delta S_i\rangle, \tag{10.2}$$

where S_0 is the nominal value for the criterion, $\langle \Delta S_i\rangle$ is the mean of the change of the criterion S, due to the error in the parameter i out of a number of j parameters. For small errors, the mean would approach the nominal performance, $\langle S\rangle = S_0$. The variance can be estimated by,

$$\sigma_S^2 = \sum_{i=1}^{j} \sigma_i^2, \tag{10.3}$$

where σ_i^2 is the variance of the change of criterion S, due to the error in the parameter i.

10.7 RSS Rule

Out of the variance, σ_S^2 follows the Root Sum Square (RSS) rule. This estimates the standard deviation of the probability density function of the criterion. By using the square of the criterion change, ΔS_i^2, due to the parameter, i, instead of the variance, σ_i^2, the RSS rule is written as,

$$\sigma_S \cong \sqrt{\sum_{i=1}^{j} \Delta S_i^2}.$$ (10.4)

The RSS rule provides the following insights. First, the statistical worst case estimate for n errors that produce the same criterion changes is $\sqrt{n}\Delta S_i$; this is not as pessimistic as adding all the changes as $n\Delta S_i$. Second, it is the large criterion changes that count much more as they enter as their squares. Thus, if we have ten parameters that produce changes of 1, and one parameter that produces a change of 10, the RSS rule indicates that the impact on the standard deviation of the former parameters is $\sqrt{10}$, while the impact of the latter parameter is $\sqrt{100}$.

The RSS rule also helps to allocate tolerance budgets to different aspects of a lens system. For example, for a diffraction limited system, the total allowed wavefront error budget might be set to 0.0707λ RMS. This budget is allocated according to the RSS rule as 0.03λ RMS for the lens design, 0.04λ RMS for the assembly, and 0.05λ RMS for the fabrication ($0.03^2 + 0.04^2 + 0.05^2 = 0.0707^2$).

10.8 Monte Carlo Simulation

In a Monte Carlo simulation the constructional parameters of a lens are chosen randomly from ranges defined by the nominal parameter values and their error probability distribution. The parameters in error are used to construct a lens trial, compensators are applied, and the system tolerancing criterion change is determined. Many Monte Carlo trials are done to determine the statistics of the tolerancing criterion change. The mean of the tolerancing criterion and its standard deviation are determined from the list of criterion changes. Depending on the application a Monte Carlo simulation may start with 100 trials to check for the appropriateness of the lens modeling, then 1,000 trials, or more. As the trials increase, it is expected that the mean and the standard deviation converge as the square root of the number of trials, $\sqrt{\#trials}$. A rule of thumb is to execute a number of trials in the order of the square of the number of parameters under error. The modeling of a lens system for tolerancing can be an art and a science, as it can be quite elaborated to properly reflect the environment, materials, fabrication and assembly errors, and more. As the lens system must be optimized for each trial using the compensators as variables, Monte Carlo simulations may take long times to run. At the end, the goal is obtaining the statistics of the as-built lens and to assign tolerances for fabrication.

Table 10.4 *Monte Carlo trials, nominal criterion 0.34λ*
RMS, mean 0.421λ RMS, standard deviation 0.047λ RMS

Trial #	Criterion	Change
1	0.441	0.101
2	0.480	0.140
3	0.369	0.029
4	0.396	0.056
5	0.445	0.104
6	0.409	0.069
7	0.390	0.050
8	0.357	0.017
9	0.516	0.175
10	0.403	0.063

Each parameter error may have its own probability density function, such as uniform, truncated normal, end-limited, and others. Once the lens trial is constructed with the parameters in error, the lens is optimized using the compensators. When lens decenters, or surface tilts, are lens errors, the lens loses its axial symmetry and, therefore, it is important to properly sample the field of view to determine correctly the tolerancing criterion such as RMS spot size, or RMS wavefront error.

10.9 Monte Carlo Simulation Example

Consider a Cooke triplet lens, as shown in Figure 10.2. The focal length is $f' =$ 50 mm, the field of view (FOV) is ±24 degrees, and the optical speed is $F/5$. The tolerances assigned are: thickness ±0.1 mm, radius ±2.5 fringes, index ±0.0005, surface figure ±0.5 fringe, and surface tilt ±1.5 arc-minutes. A truncated normal distribution for these errors is assumed. The field of view is sampled at the field center and four full field positions. The back focal length was used as a compensator. A lens decenter can be decomposed as two surface tilts and a thickness change. However, for small surface tilts the thickness change is negligible. Thus, for simplicity and clarity, here only surface tilts in two directions are allowed.

Table 10.4 shows the results of ten Monte Carlo runs, which give a mean value of 0.421λ RMS, and a standard deviation of 0.047λ RMS. The nominal wavefront error is 0.34λ RMS. Depending on the performance requirements, on the parameters that most degrade the tolerancing criterion, the optics shop's ability to meet tolerances, and cost, the tolerances can be made tighter or looser

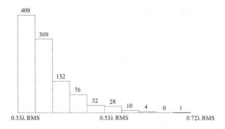

Figure 10.4 Histogram of 1,000 Monte Carlo runs for a Cooke triplet lens.

to meet the requirements and cost. This is a simple example to illustrate how the tolerancing criterion statistics are obtained. However, for a lens to be manufactured the tolerancing process is often elaborated to properly model the as-built lens.

Figure 10.4 shows the histogram of 1,000 Monte Carlo trials for the same Cooke triplet lens, the mean value is 0.4λ RMS, and the standard deviation is 0.055λ RMS. Each histogram bin has trials with performance within about 0.04λ RMS. Thus, there are 408 lens trials with a tolerancing criterion between 0.33λ RMS and 0.37λ RMS. Therefore, under the tolerances specified there is a percentage probability of about 40.8% that the lenses will perform within 11.8% from the nominal performance. Also, there is a probability of 71.7% that the lenses will perform within 23.6% from the nominal performance. If uniform distributions are chosen for the parameters then the mean would be 0.42λ RMS and the standard deviation would be 0.077λ RMS. Thus, properly modeling the parameters error distribution provides a more accurate level of tolerancing.

Because the errors in the fabrication of a lens can substantially degrade the lens performance, it is important to minimize as much as possible the nominal tolerancing criterion during the lens optimization, so that there is more room to accommodate for such errors. However, different forms of optical systems that satisfy the requirements for an application may have more or less sensitivity to fabrication errors for the same level of nominal image quality.

Table 10.5 provides the mean and standard deviation when 1,000 trials at a time were performed for a given category of error. The change in the mean of the tolerancing criterion for errors in thickness, as well as its standard deviation, are large. Clearly, and by far, the worst offender is the category of thickness errors. Tightening the tolerance in thickness will be a choice. For example, by decreasing the thickness tolerance to ±0.05 mm, the criterion mean would be 0.34λ RMS, and the standard deviation would be 0.016λ RMS. This would make 81% of the lenses perform within 10% of the nominal

Table 10.5 *Cooke triplet lens. Mean and standard deviation for categories of errors, nominal mean 0.328λ RMS*

Parameter category	Mean λ RMS	Standard deviation λ RMS
Radius	0.329	0.0038
Thickness	0.378	0.0520
Surface tilt	0.334	0.0056
Figure	0.328	0.0033
Index	0.328	0.0022

Table 10.6 *Constructional data of the Cooke triplet lens, f = 50 mm, FOV = ±24°, F/5*

Surface	Radius (mm)	Thickness (mm)	Glass
1	26.6335	3.25	N-LAK33
2	426.1623	6.0	
3 (Stop)	−25.9915	1.0	TIF6
4	25.0718	4.75	
5	169.8704	3.0	N-LAK33
6	−23.2263	42.3551	

criterion. However, lens manufacturers put a cost premium on tight tolerances for thickness because of the risk of over-grinding the lens and the need to start over with a new blank lens. A next step would be to explore using the airspaces as compensators to avoid tightening the lens thickness tolerance. This might result in a tedious and costly lens assembly. Table 10.6 provides the constructional data of the Cooke triplet. Thus, increasing the lens production yield is most often a trade-off with cost.

10.10 Behavior of a Lens under Manufacturing Errors

Under fabrication errors, that is under lens perturbation, a lens system suffers from a number of optical effects. These can be divided as relating to axial symmetry and not relating to axial symmetry. Errors in radii of curvature, lens thickness, and index of refraction maintain the axial symmetry of a lens. Errors in surface tilt break the axial symmetry.

The first-order effects to take place are that the focal length changes, and that the image is displaced laterally. This lateral image displacement is known as bore-sight error, or line of sight error, and arises from the lenses becoming

Table 10.7 *Changes that take place under perturbation according to symmetry and to aberration order*

Changes that relate to axial symmetry		Changes that relate to lack of axial symmetry	
First-order	Aberration	First-order	Aberration
Focal length	Spherical aberration	Image lateral displacement	Uniform coma
Image size	Linear coma aberration	Anamorphic image distortion	Uniform astigmatism
		Chromatic change of line of sight	Linear astigmatism
			Field tilt

wedged. In addition, for each wavelength, the image displacement might be different. The second effects that take place are changes in the aberrations, and that new aberrations appear. Table 10.7 provides a summary of these effects according to the symmetry, and whether they are of first-order, or relate to aberrations. In the same way that spherical aberration W_{040} is uniform over the field of view, uniform coma and uniform astigmatism can now be present over the field of view. Linear coma W_{131} grows linearly with the field of view; now linear astigmatism and linear focus, this is field tilt, can take place.

The change in focal length of a thin lens is given by,

$$\Delta f = \frac{\Delta n}{n-1} f. \tag{10.5}$$

A change in the index of refraction of 0.001 results in a change of focal length of approximately 0.2%. Index of refraction can be measured to 1×10^{-5}, and is usually sufficiently well known. Thus, system changes due to errors in the index of refraction are expected to be very small. However, for diffraction limited systems it is important to check the index of refraction of the materials being used. The index of refraction homogeneity is also of concern, as a difference in index of 0.0001 in a 10 mm glass blank produces an optical path difference of 0.001 mm, or about two wavelengths in the visible spectrum.

Having an understanding of the effects that take place when a lens is perturbed can allow a lens designer to control, or mitigate, them to avoid specifying tight tolerances. For example, the tilt of an image sensor can be used to match field tilt aberration, or some radial adjusting screws can be designed into a lens barrel to laterally displace a lens and correct for uniform

Table 10.8 *Lens configuration setting for desensitizing a Cooke triplet lens for lens element and tilt errors*

Configuration	1	2	3	4	5	6	7
Focal length, mm	100						
Lens #1 decenter, mm		0.025					
Lens #2 decenter, mm			0.025				
Lens #3 decenter, mm				0.025			
Lens #1 tilt					0.05°		
Lens #2 tilt						0.05°	
Lens #3 tilt							0.05°

coma. This has been done in adjusting microscope objectives. Alternatively, a lens airspace can be adjusted to correct for residual spherical aberration.

Uniform astigmatism depends on the square of the surface tilt. Since the lens tilts under consideration are small, uniform astigmatism is negligible. Thus, if this aberration is detected in a nominally axially symmetric lens system, it is likely due to surface figure error or to a lens being deformed due to improper mounting. Table 1 in Appendix 4 summarizes the primary aberrations that can take place in a plane symmetric system.

10.11 Desensitizing a Lens from Element Decenter, Tilt, or Wedge

Lens design programs allow us to set multi-configurations for a lens system. Each configuration may differ, for example, in constructional parameters, in field of view, in relative aperture, and in wavelength choice. An opto-mechanical engineer is concerned about lens decenter and tilt tolerances. To desensitize a lens, say a Cooke triplet lens, seven configurations are defined. One configuration is the nominal configuration; three configurations are for lens element decenter, one for each lens; and three configurations are for lens element tilt, one for each lens; this setting is shown in Table 10.8. The error function for the nominal configuration has the first-order lens constraints and may include image quality performance. The error function for the remaining six configurations has only image quality performance.

Lens decenters and lens tilts are set only in one direction, so as to reduce the lens system symmetry to plane symmetry. The field of view needs to be properly sampled, as there is no longer axial symmetry for six configurations. However, because the system becomes plane symmetric and the main effects

Table 10.9 *Lens configuration setting for desensitizing a Cooke triplet lens for lens wedge*

Configuration	1	2	3	4	5	6	7
Focal length, mm	100						
Surface #1 tilt		0.1°					
Surface #2 tilt			0.1°				
Surface #3 tilt				0.1°			
Surface #4 tilt					0.1°		
Surface #5 tilt						0.1°	
Surface #6 tilt							0.1°

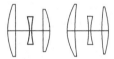

Figure 10.5 Desensitized Cooke triplet lens. Left, standard lens solution; Right, desensitized lens solution. The front positive lens takes a meniscus form, and the rear positive lens takes a double convex form. Glasses are N-LAK33, TIF6, and N-LAK33. FOV = ±24° at *F*/5.

of surface tilt are uniform coma and linear astigmatism, sampling three or five fields in the plane of symmetry might be sufficient. Then optimizing such a multi configuration lens system will tend to desensitize the lens system for lens element decenter and tilt errors. Performance gains of 5%, 10%, or more are often obtained. If the desensitizing is not sufficient, then a different lens solution is desensitized and evaluated until the specified lens yield is achieved for a given set of tolerances. In this desensitizing method, the lens optimizer will change the lens to mitigate the worst offenders to performance first, due to lens element decenter and tilt errors. Compensators can be added by releasing as variables appropriate lens parameters, such as the back focal length, the image plane tilt, or a lens decenter.

For spherical surfaces a lens decenter can be resolved as two surface tilts and a thickness change; however, the thickness change is negligible.

The optics shop is concerned about lens wedge tolerances. Lens system desensitizing for a lens wedge can be done by tilting each of the surfaces of a lens system in a multi-configuration, as shown in Table 10.9 for a Cooke triplet lens. Then optimization under surface tilt perturbation may provide a desensitized lens. Figure 10.5 shows the form of a Cooke triplet lens that has been desensitized to surface tilt. The main offender to the performance of the Cooke triplet was sensitivity to linear astigmatism, which was mitigated. In this case it was necessary to set each surface tilt first to 1.0° and then to 0.1° for the optimizer to escape from a standard solution and find the less sensitive lens solution.

10.12 Lens Drawings

Once a lens is designed and tolerances have been assigned for manufacturing, lens drawings need to be produced. A lens drawing should provide sufficient information so that the correct lens is made. It is important to have effective communication with the lens manufacturer to reflect the shop fabrication skills and to avoid mistakes.

Further, it is imperative that the lens designer checks, and double checks, a lens design to make sure there are no errors. A lens designer needs to also have effective communication with the opto-mechanical engineer who will design the lens barrel, and the lens assembly engineer, to make sure the lens can be made and be assembled and aligned.

There are a variety of lens formats for lens drawings according to organization or company. However, a sample of the International Organization for Standardization (ISO) standard for drawings for optical elements and systems, ISO 10110-10, is shown in Figure 10.6.

In addition to the lens drawing in the top of Figure 10.6, the three boxes in the bottom are for providing the following information, for each surface and for the material: Material type, index of refraction, and ν-number; Radius of curvature, convex CX, concave CC, and tolerance; Clear aperture or optically effective diameter; Protective chamfer; Surface treatments and coatings; Stress

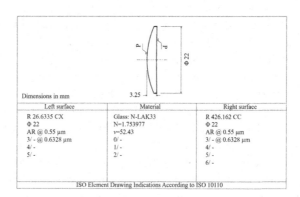

Dimensions in mm		
Left surface	Material	Right surface
R 26.6335 CX	Glass: N-LAK33	R 426.162 CC
Φ 22	N=1.753977	Φ 22
AR @ 0.55 μm	ν=52.43	AR @ 0.55 μm
3/ - @ 0.6328 μm	0/ -	3/ - @ 0.6328 μm
4/ -	1/ -	4/ -
5/ -	2/ -	5/ -
		6/ -
	ISO Element Drawing Indications According to ISO 10110	

Figure 10.6 Lens drawing according to ISO 10110-10.

birefringence; Permissible bubbles and other glass inclusions; Glass homogeneity; Surface irregularity; Centering tolerance, or wedge; Surface imperfections allowance; Laser damage threshold indication; and Whether the surface is cemented or optically contacted.

Further Reading

Bates, R. "Performance and tolerance sensitivity optimization of highly aspheric miniature camera lenses," *Proceedings of SPIE*, 7793 (2010), doi: 10.1117/12.860919.

Bauman, Brian J., Schneider, Michael D. "Design of optical systems that maximize as-built performance using tolerance/compensator-informed optimization," *Optics Express*, 26 (2018), 13819–40.

Fuse, K. Method for designing a refractive or reflective optical system and method for designing a diffraction optical element, USP 6,567,226 (2003).

Grey, D. S. "Tolerance sensitivity and optimization," *Applied Optics*, 9(3) (1970), 523–26.

Herman, E., Sasián, J. "Aberration considerations in lens tolerancing," *Applied Optics*, 53 (2014), 341–46.

Rimmer, M. "Analysis of perturbed lens systems," *Applied Optics*, 9(3), (1970), 533–37.

Rogers, J. "Using global synthesis to find tolerance-insensitive design forms," Proceedings of SPIE 6342, International Optical Design Conference 2006, 63420M (2006), doi: 10.1117/12.692251.

Rogers, J. "Global optimization and desensitization," Proceedings of SPIE 9633, Optifab 2015, 96330S (2015), doi: 10.1117/12.2196010.

Sasián, J., McCormick, F. B., Webb, R., Crisci, R. J., Mersereau, K. O., Stawicki, R. P. "Design, assembly, and testing of an objective lens for a free-space photonic switching system," *Optical Engineering*, 32(8) (1993), 1871–78.

Schwiegerling, J. *Optical: Specification, Fabrication, and Testing* (Bellingham, WA: SPIE Press, 2014).

Smith, W. J. "Fundamentals of establishing an optical tolerance budget," *Proceedings of SPIE, Geometrical Optics*, 0531 (1905), 196–204.

Weichuan, Gao, Youngworth, Richard N., Sasián, José. "Method to evaluate surface figure error budget for optical systems," *Optical Engineering*, 57(10) (2018), 105108.

Youngworth, R. N. "Statistical truths of tolerance assignment in optical design," *Proceedings of SPIE*, 8131 (2011), 81310E.

11

Using Lens Design Software

There are a variety of lens design computer programs. Some of them are CODE V, OpTaliX, OpticStudio, Oslo, and Synopsys. The costs of these programs are two-fold, one for the license to use them, the other for the time the engineer must spend learning to use them, which can be substantial.

Lens design software allows for the setting and analysis of a lens or lens system, provides a variety of lens evaluation tools, permits the optimization of a lens, allows the tolerancing of a lens system, and provides a variety of other useful tools.

To acquire the skill of lens design one must become familiar with at least one design program, develop some proficiency in its use, and become confident on the calculations produced by the program. Lens design software is not necessarily correct, and a lens designer's task is to double-check the results of the program. On the other hand, the lens design program is expected to be correct and the lens designer can be wrong on a given lens design issue. Then there is the opportunity to learn.

There must be a turning point when the user is comfortable, and enjoys using a lens design program. The user becomes excited about using the software because current lens design programs are quite powerful in what they can model, analyze, and optimize. Ultimately, lens design programs allow the user to better understand optical systems, to provide solutions to optical engineering problems, and to exercise creativity. These aspects make lens design software quite appealing as professional tools.

A good way to start using a lens design program is asking for a demo and help by someone who knows how to use the program. This may avoid frustration, which may discourage an engineer from going forward in developing the skill of lens design. Lens design software companies offer short courses on how to use their software.

Using a lens design program is not about randomly, or unintelligently, selecting optimization options, but intelligently using the program. This requires planning and visualizing how the lens design problem will be tackled, pausing, reflecting on the data the program displays, and rethinking the way to proceed. Lens designers often optimize a lens system to extract the last bit of good performance that the lens is capable of, given as a set of constraints such as materials, packaging, and cost.

This chapter provides a discussion about using a lens design program, with emphasis on lens optimization.

11.1 Utilities and Settings

Once the data of a lens has been input into a lens design program, there are a variety of utilities and settings to help analyze, adjust, or optimize the lens. These can help to save time and to decrease the computing overhead of the lens optimizer, and to properly conduct a lens analysis and optimization.

Surface pickups refer to setting a given surface parameter, from a parameter, not necessarily of the same type, of a preceding surface. If the lens is symmetrical about the stop, then using pickups to set the rear part of the lens by copying, this is by picking up, the parameters of the front part would be a choice.

Solves provide a surface curvature or a thickness using first-order calculations. A thickness solve will provide the distance to the next surface for a given ray height at that next surface. A thickness solve is often used to find the first-order distance to the ideal image plane. A curvature solve can set the marginal ray slope equal to zero in image space, making the image be located at infinity, or the lens afocal if the object is at infinity. Solves may reduce the computational burden on the lens design optimization algorithm and save time.

Macros are programs that the lens designer writes to be executed by the lens design program. Macros allow lens designers to implement customized calculations not provided by the lens design program as standard options, and they are a powerful tool.

Variables are used by the optimization algorithm, or optimizer, to improve the lens. Any constructional parameter of the lens can be a variable. The number of lenses, lens element optical power, lens bending, lens separation, lens surface asphericity, and lens material are variables frequently used in lens optimization.

Weights are numerical values given to field points, wavelengths, and lens configurations, to favor/disfavor their contribution in a given analysis, or to aberrations during lens optimization.

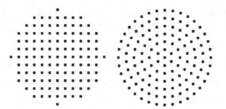

Figure 11.1 Grids to define rays for analysis and optimization.

For the results of a program to properly reflect the status of correction of a given lens system, it is necessary that proper sampling of the field of view, the aperture, and the spectral bandwidth be defined for analysis and optimization. If there are a few field points defined, then it is possible that the performance of the lens would be acceptable for those points, but not acceptable for other field points not included in the analysis or optimization. For axially symmetric systems, three field points can be set to start an optimization, for example, $H = 0$, $H = \sqrt{2}/2$, and $H = 1$ are traditional field points. The number of field points can be increased according to the extent of the field of view, and to the number of aspheric surfaces in the system; it is not unusual to define twelve field points. Field points can also be set to define circular regions covering equal areas of the field of view.

If too few rays are defined to sample the pupil, then computations like RMS spot size, RMS wavefront error, or MTF might be in meaningful error. Figure 11.1 shows two ray grids, square and polar, that can be used to define ray distributions at a pupil. For optimization of complex systems that are intended to be diffraction-limited, and where computing time is important, other more efficient grids are used.

To properly evaluate the point spread function, or any polychromatic system performance metric, it is necessary to consider the spectral sensitivity of the light sensor to be used, and the light transmission of the lens system, by including enough wavelengths with appropriate weighting. For example, for a visual system the design wavelengths can be given a weight based on the spectral sensitivity of the eye, shown in Figure 11.2 and given in Table 11.1. Figure 11.3 shows the relative sensitivity of the eye for photopic and scotopic vision in logarithm scale vs. the wavelength, and by scaling up the scotopic vision values to reflect that scotopic vision is sensitive to very low light levels.

Thus, a lens designer is concerned with setting the sampling options in the different menus of a lens design program, to provide accurate results, and as quickly as possible, to save time. One way to determine sampling size for a given calculation is to start with the minimum sampling value and note the

Table 11.1 *Normalized spectral sensitivity of a human eye for photopic (P)*
and scotopic (S) vision

λ nm	P	S	λ nm	P	S	λ nm	P	S
400	0.00	0.01	500	0.32	0.98	600	0.63	0.03
410	0.00	0.03	510	0.50	0.99	610	0.50	0.01
420	0.00	0.09	520	0.71	0.93	620	0.38	0.01
430	0.01	0.20	530	0.86	0.81	630	0.27	0.00
440	0.02	0.33	540	0.95	0.65	640	0.18	0.00
450	0.04	0.46	550	0.99	0.48	650	0.11	0.00
460	0.06	0.57	560	0.99	0.33	660	0.06	0.00
470	0.09	0.69	570	0.95	0.20	670	0.03	0.00
480	0.13	0.79	580	0.87	0.12	680	0.02	0.00
490	0.21	0.90	590	0.76	0.07	690	0.01	0.00

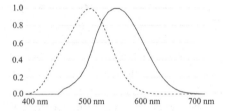

Figure 11.2 Normalized spectral sensitivity of the human eye for photopic vision
(solid line) and for scotopic vision (dashed line). The peak for photopic vision is at
about 555 nm, and for scotopic vision is at about 507 nm.

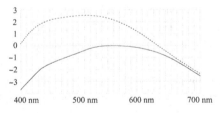

Figure 11.3 Photopic (solid line) and scotopic (dashed line) vision in logarithm
scale vs. the wavelength of light.

results of that calculation. Then proceed to the next sampling value and note
the results of the calculation, and proceed, increasing the sampling, until there
is no meaningful change in the results of the calculation.

11.2 Merit Function

In order for a lens to be evaluated or improved, a merit function *MF*, also
called error function, must be constructed. This function conveys

specifications of the lens system and serves as the input for the lens design optimization algorithm. The merit function can be written in generic form as,

$$MF^2 = \frac{\sum_{i=1}^{n} W_i^2 (T_i - A_i)^2}{\sum_{i=1}^{n} W_i^2}, \qquad (11.1)$$

where n is the number of lens performance metrics, A_i, that quantify the system performance, T_i is the target value for metric A_i, and W_i is the weight given to that metric i. For example, A_i can be a ray intercept error, the RMS wavefront error for a given field point, distortion aberration, the focal length, or any other measurable aspect of a lens. The merit function can also be constructed as,

$$MF^2 = \frac{W_{iq}^2 \cdot MF_{iq}^2 + W_{pkg}^2 \cdot MF_{pkg}^2 + W_{dist}^2 \cdot MF_{dist}^2 + W_{fo}^2 \cdot MF_{first-order}^2}{W_{iq}^2 + W_{pkg}^2 + W_{dist}^2 + W_{fo}^2}. \qquad (11.2)$$

This construction is a sum of sub-merit functions for the categories of image quality, packaging constraints, distortion aberration, first-order constraints, and/or other lens performance metrics. This form allows us to conduct category trade-offs of performance through the weights, W_i, of the different categories.

It is critical to be thoughtful on the merit function construction, as an improper function is likely to yield improper results. Usually constructing the merit function by constraining the focal length to a given value, and by including the RMS spot size, or RMS wavefront error, is sufficient to optimize many lens systems. If packaging problems arise during the optimization, they can be dealt with, one by one, when they appear, by adding constraints to control them. For example, the minimum central thickness of a lens, or the minimum air space, are typical constraints to add to the merit function.

Through weight selection, W_i, two items can be directed: one is the order in which the optimization algorithm will address each performance metric, A_i, also called the optimization operand, and the other is how well the metrics will be optimized, or balanced, relative to each other. It is good practice to keep the merit function as simple as possible to reduce computing time, to allow the optimizer to find solutions, and to avoid conflicting requirements between optimization operands. For example, optimizing simultaneously using aberration coefficients and RMS spot size in the merit function construction is often conflicting. Other optimization problems can result from releasing too many variables; for example, too many aspheric coefficients. Sometimes when the radius of curvature of a surface is large, the conic constant may take very large values and can make the optimization unstable. In this case the conic constant is frozen, so no longer set as a variable, and perhaps the fourth-order aspheric coefficient is released instead as a variable.

11.3 Optimization Algorithm

There are two types of optimization algorithms, namely, local optimization and global search. A goal of an optimization algorithm, or optimizer, is to minimize the merit function. It is an automatic process once it is set and started. Some optimization algorithms are: Damped Least Squares (DLS), Simulating Annealing, Genetic Algorithms, Simplex Method, and Orthogonal Descent. Perhaps the most used and proven is the DLS algorithm. In DLS, derivatives of the merit function, as a function of the optimization variables, are computed, and changes of the constructional parameters are determined that may decrease the error function. It is an iterative process until a given condition, such as how small the merit function is, or the number of iterations, is reached. In minimizing the error function, a plurality of local performance minima can be reached, and some other higher or lower minima may not be reached. The optimization algorithms may suffer the common problem of stagnation in which they stay on a local minimum without being capable of escaping to a lower minimum. A task of the lens designer is to help the optimizer finding a solution near, or at, the global minimum to a given lens design problem.

The lens designer chooses the parameters, or variables, that the optimizer will use to decrease the merit function. The goal is to allow the optimizer to vary lens constructional parameters that effectively lead to a decrease in the merit function. Lens curvatures and aspheric coefficients are usually effective variables. Lens thickness is usually not an effective variable but in some cases it can be. For example, a thick meniscus lens can be used to correct for Petzval field curvature aberration. Aberration theory provides insight into what can be an effective optimization variable. In addition, some physical constraints can be set as part of the merit function, such as the focal length or the system total track length. These constraints can be set as soft or hard. According to weight, soft constraints may not be met perfectly, while hard constraints will be met exactly by the optimizer.

There are also global search algorithms whose function is to find different solutions and hopefully the global minimum to a given lens design problem defined by specifications and constraints.

11.4 Analyzing a Lens

The constructional data of a lens, the field of view, the spectral bandwidth, and the $F/\#$ can be entered into a lens design program for analysis purposes. This analysis, among many others, can include image quality assessment, determining lens dimensions, determining the position of the pupils, determining light transmission, or assessing how temperature changes affect the lens performance.

An analysis can also be conducted on combinations of several constituent lens systems that form a complete system. For example, endoscope systems may comprise a first objective lens, then one or more optical relays systems, and a final imaging lens.

A problem that arises in lens design is that lens data can be incorrectly input into the lens design program and errors can take place. For this reason, checking that the lens program data editor or manager has correct information is a task that lens designers often do. Sometimes the data input has been done correctly but the data is faulty, and then it must be corrected.

11.5 Adjusting a Lens

A given lens that has been entered into a lens design program can be adjusted for a variety of reasons. For example, the scale of a lens can be changed by using the lens design program scaling function. Lens scaling makes it easy to meet a given focal length or packaging requirement. While maintaining the individual lens element optical powers, the surface curvatures can be varied to adjust the lens for any data variation such as glass index of refraction, slight errors in curvatures, or a different operating temperature. The aspheric surface coefficients can also be adjusted to reflect any departure in lens use from the nominal lens specification. It is not uncommon to adjust aspheric coefficients to correct errors in the specified coefficients.

A lens can be adjusted to reflect the actual glass available for fabrication, or to fit the surface curvatures to a manufacturer's test plates list, or to meet the lens thickness requirements needed for lens element mounting and fabrication.

A lens can be adjusted for a different field of view by trading-off its optical speed. For example, a Cooke triplet lens can be nominally designed for a half-field of view of 26° at $F/5.6$, and later adjusted for a HFOV of 31° at $F/8$, 16° at $F/4$, or 7° at $F/2.8$, while maintaining the same image quality. This adjustment requires setting a merit function such as RMS spot size and re-optimizing the lens. The lens thicknesses and diameters would be also adjusted to reflect the change in light passage through the lens, and to avoid larger and thicker lenses than necessary.

11.6 Modifying and Improving a Lens

There are many reasons to modify a lens, such as improving image quality, improving resolving power and image brightness by decreasing the $F/\#$, improving relative illumination, increasing the field of view, changing the stop position to satisfy first-order requirements such as telecentricity or to make the lens more or less symmetrical about the stop, reducing tolerances, or reducing lens size, weight, and volume. Some well-known techniques for modifying and improving a lens are:

1. Determining the limiting aberrations from wave fans, ray fans, or aberration coefficients, and implementing effective design variables to correct, balance, or mitigate those aberrations.
2. Scaling a lens.
3. Shifting the stop aperture.
4. Splitting a lens element optical power into two individually weaker lens elements with the same total power as the un-split lens.
5. Bending a lens element.
6. Removing a lens element that has a relatively long focal length.
7. Allowing light vignetting.
8. Making one or more surfaces aspheric.
9. Reversing a lens element, a doublet, a group of lens elements, or the entire lens system and re-optimizing.
10. Interchanging groups of lenses from different lens systems. For example, some lens systems can be analyzed as having a front and a rear group. Then the front and rear parts of two lens systems can be interchanged to create four lens systems. Usually one lens system will perform best for a given application.
11. Temporarily setting a lens element as aspheric to find out if the lens system improves. If so, then the aspheric lens can be replaced often by a doublet lens with spherical surfaces.
12. Introducing in an airspace an aspheric optical element with very small, or none at all, optical power and re-optimizing the lens.
13. Introducing a nearly concentric meniscus lens.
14. Increasing the index of refraction of the lens elements.
15. Selecting different lens materials.
16. Increasing the glass v-number difference between elements where the marginal ray height is the largest, to help correct chromatic aberration and reduce lens curvatures.
17. Adding a field flattener lens.
18. Including diffractive optical elements, or gradient index materials.

19. Reducing the RMS of the marginal ray angle of incidence in the lens surfaces.
20. Trading-off image quality and packaging requirements such as lens system length.

It is important to keep in mind that large changes to a lens may result in the optimization algorithm failing to find a local minimum and the error function diverging. Thus, during lens optimization the designer guides the optimizer to find a minimum, and also helps the optimizer to walk through a number of minima until an acceptable solution is reached. It is good practice to start an optimization with a lens that is corrected for some of its primary aberrations, so as to help the optimizer to find a merit function minimum. Then the lens designer modifies the lens with small manual changes, and re-optimizes the lens using the optimizer. This is repeated until a lens solution is reached. Finding a lens solution and modifying it many, many times with the help of the optimizer is typical of an optimization session. Tens or hundreds of lens files can be generated in a few optimization sessions.

Global search optimizers that use artificial intelligence may require minimum attention from a lens designer to find solutions, though this does not mean that such optimizers can automatically design lenses that solve a given lens design problem. Intelligent and creative input from a lens designer is indispensable for the design of state-of-the art lens systems.

11.7 Designing a Lens

A lens system can be designed from first principles by adding complexity to a simple lens. This requires knowledge of first-order optics, aberration theory, and lens design experience.

Alternatively, an existing lens from a data base or from the patent literature can be used as a starting point. In this case the lens is first adjusted to meet the necessary focal length, field of view, and f-number required. Then the lens is optimized to find out how good its image quality can be, and further modified until all the lens specifications are met. New materials can help to push the level of performance of a lens system.

A lens design can also be carried out by combining different lens groups, or parts, from different lens systems, or by combining or concatenating well corrected lens systems.

A global search optimizer can be used to find many solutions to be analyzed, adjusted, and fine optimized, including lens tolerancing considerations.

11.8 Inventing a New Lens

Lens designers adjust and modify an existing lens on a routine basis to improve it or set it for fabrication. This may not be considered as inventing a new lens, unless the modifications rendered a substantial level of performance improvement with respect to prior art. Clearly differentiating why a lens performance is meaningfully superior, and showing why the lens is novel, and not merely a trivial modification of an existing lens form, helps to support it as an invention.

A new lens form can evolve from existing lens forms by adding incremental improvements over time, and taking advantage of new materials. Such is the case of miniature lenses for mobile phones or lenses for micro-lithography. Occasionally, new applications, or demands, result in lens forms that have not been explored in the past, or in lenses that meaningfully exceed the performance of previous lens generations. The Petzval portrait objective was innovative because of its superior image quality and fast optical speed. It is said that the commercial success of the Petzval portrait lens was immediate and extraordinary, and that it spread with unexpected rapidity.

There is plenty of opportunity to invent new useful lens systems. This requires having lens design training, understating new demands and applications, being aware of new materials, being creative, and seizing opportunities.

11.9 Documenting a Lens

Once a lens design is considered completed, it is appropriate to document the design, organize the lens files with useful file names that include the project name and date of creation, and properly archive the documentation and lens files. One compelling reason to create this documentation is to save time in the future, as thousands of lens design files can be created over a year.

Further Reading

Dilworth, D. C., Shafer, D. "Man versus machine: a lens design challenge," *Proceedings of SPIE*, 8841 (2013), 88410G.

Feder, D. P. "Automatic optical design," *Applied Optics*, 2(12) (1963), 1209–26.

Forbes, G. "Optical system assessment for design: numerical ray tracing in the Gaussian pupil," *Journal of the Optical Society of America A*, 5(11) (1988), 1943–56.

Kingslake, R., Johnson, B. "Automatic lens improvement programs," Chapter 17, in *Lens Design Fundamentals* (Burlington, MA: Academic Press, 2010).

Merion, J. "Damped least-squares method for automatic lens design," *Journal of the Optical Society of America*, 55 (1965), 1105.

Vasiljevic, D. *Classical and Evolutionary Algorithms in the Optimization of Optical Systems* (Boston: Kluwer Academic Publishers, 2002).

Yabe, Akira. *Optimization in Lens Design* (Bellingham, WA: SPIE Press, 2018).

Link to Lambda Research Corporation supplier for OSLO optical design software (2018); https://www.lambdares.com/oslo-usonly/

Link to Optenso™ supplier for OpTaliX optical engineering software (2018); http://www.optenso.com/links/links.html

Link to Optical Systems Design, Inc. supplier for Synopsis™ lens design software (2018); http://www.osdoptics.com/

Link to Synopsys®, Inc., Optical Design Solutions, supplier for CODE V lens design software (2018); https://www.synopsys.com/optical-solutions.html

Link to Zemax, supplier of OpticStudio®, optical design software (2018); https://www.zemax.com/products/opticstudio/

12

Petzval Portrait Objective, Cooke Triplet, and Double Gauss Lens

Three well-known and important classical lens forms are the Petzval objective, the Cooke triplet lens, and the double Gauss lens. An understanding about how these lens forms work, and how they are designed, provides a solid background to push forward the skill of lens design. Many other lens forms are derived from such classical lens forms by lens splitting and adding lens complexity.

In this chapter we discuss the Petzval sum, the concept of stress and relaxation in lens design, and those classical lenses.

12.1 Petzval Sum

The Petzval sum provides the vertex radius of curvature of the surface where an image falls whenever there is no astigmatism aberration. The Petzval sum is

$$\frac{1}{n_k'\rho_k'} - \frac{1}{n_1\rho_1} = -\sum_{i=1}^{k} \frac{n_i' - n_i}{n_i n_i' r_i}, \tag{12.1}$$

where ρ_1 is the vertex radius of the object surface, $\rho_{k'}$ is the Petzval radius, n is the index of refraction, and k is the number of surfaces in the lens system. If the Petzval radius is negative (positive), the center of curvature of the Petzval surface is to the left (right) of the surface vertex. Notably, the Petzval sum does not depend on lens thicknesses. For a system of thin lenses in air, and for a flat object, the Petzval sum simplifies to,

$$\frac{1}{\rho_{Petzval}} = -\sum_{i=1}^{k} \frac{\phi_i}{n_i}, \tag{12.2}$$

137

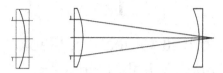

Figure 12.1 Left, a parallel glass plate made out of a plano convex and a plano concave lens in contact. Right, lens separation results in optical power, but not in Petzval field curvature.

where ϕ_i is the optical power of the i thin lens. A thin lens of 100 mm focal length made out of BK7 glass has a Petzval radius of $\rho_{Petzval} = -151.7$ mm.

If we assume that a lens system in air is made out of the same glass, $\rho_1 = \infty$, multiply the Petzval sum by the square of a given height, h, above the optical axis, and divide by two, one obtains,

$$\frac{h^2}{2\rho_{Petzval}} = -\frac{n'-n}{nn'}\sum_{i=1}^{k}\frac{h^2}{2r_i} = -\frac{n'-n}{nn'}Sag, \qquad (12.3)$$

where Sag stands for the cumulative sag of all the optical surfaces in the lens system at the height h. This result indicates that, to have a flat field lens, there must be no variation of thickness across the lens aperture.

Since the image displacement, Δs, by a parallel plate of thickness t is,

$$\Delta s = \frac{n-1}{n}t, \qquad (12.4)$$

Petzval field curvature can be interpreted as the image displacement caused by a plate of varying thickness, Sag, across the lens aperture.

Figure 12.1 (left) shows a parallel plate made out of a plano convex lens and a plano concave lens, of the same but opposite optical power. The plate has no optical power and has an infinite Petzval radius. When the lenses are separated as in Figure 12.1 (right), the combination acquires optical power but maintains an infinite Petzval radius. The positive Petzval field curvature contributed by the positive lens is corrected by the contribution from the negative lens. When a lens is located near an image and corrects for Petzval field curvature aberration, it is called a field flattener lens. In this case the marginal ray height is nearly zero, and, therefore a field flattener lens does not contribute any significant amount of spherical aberration, coma, or astigmatism.

Since the Petzval sum does not depend on lens thickness, a zero optical power lens in the shape of a thick meniscus lens can have a positive Petzval radius, to balance the contributions to the Petzval sum from lenses that have a negative Petzval radius. For example, as shown in Figure 12.2, a meniscus lens made out of BK7 glass with radii of curvature of 50 mm and 43.19 mm, and a

Figure 12.2 An afocal thick meniscus lens has a positive Petzval radius.

thickness of 20 mm, is afocal, but has a Petzval radius of $\rho_{Petzval} = +930$ mm. This positive Petzval radius can be made smaller by decreasing the surface radii of curvature, or by giving the meniscus lens a negative optical power.

Aspheric optical surfaces that do not have a quadratic term as a function of the aperture, do not contribute to the Petzval sum, because they do not contribute optical power at their vertex. However, aspheric coefficients of an order higher than second influence high orders of field curvature.

There are four classic ways to control field curvature aberration. First, using high index glass for the positive lenses, and low index glass for the negative lenses to minimize and maximize, respectively, their respective contributions to the Petzval sum. Second, the use of a field flattener lens. Third, the use of a thick meniscus lens. Fourth, creating beam constrictions where the marginal ray height, y, is small and beam bulges where the marginal ray height, y, is large.

12.2 Lens Stress and Relaxation

When optical power in a lens system is obtained by the combination of positive lenses, the change of slope of the marginal ray from lens to lens is gradual and distributed among the lenses, and the lens is said to be relaxed. When the same amount of optical power is obtained by the combination of positive and negative lenses, the change of slope of the marginal can be increased from lens to lens, and in this case the lens is said to be stressed. The weighted power, w, of a surface is defined as,

$$w = \phi y = n'u' - nu, \qquad (12.5)$$

where ϕ is the optical power of the surface, y is the first-order marginal ray height at that surface, and u and u' are the ray slopes before and after refraction, respectively. The weighted power, w, is then a measure of the change of slope of a ray as it is refracted or reflected by a surface or a lens.

The structural aberration coefficients for a thin lens in air with the stop aperture at the lens, and for spherical aberration, coma, and astigmatism are proportional to w^3, w^2, and w, respectively. Thus, the weighted power is a measure of aberration and of stress in a lens. If the optical power of a thin singlet lens is split into two singlet lenses, each with half the optical power, then spherical aberration and coma of the combination decreases. This is, the lens system is relaxed.

For a system of k surfaces, a parameter, W, that quantifies how the optical power is distributed in a lens can be written as,

$$W = \sqrt{\frac{1}{k(1-m)^2} \sum_{i=1}^{k} \left(\frac{\phi y}{n_k' u_k'}\right)_i^2}, \qquad (12.6)$$

where the inclusion of the factor, $n_k' u_k'$, and the transverse magnification, m, makes W independent of the optical conjugate at which the lens system works, and independent of scale.

When a surface is not concentric to the stop aperture, $\bar{A} \neq 0$, or the pupils, or when it is not aplanatic, $\Delta(u/n) \neq 0$, field aberrations are contributed. In a lens system that has some symmetry about the aperture stop, the odd aberrations tend to cancel. A parameter, S, that quantifies symmetry in a lens can then be written as,

$$S = \sqrt{\frac{1}{k(1-m)^2} \sum_{i=1}^{k} \left(\frac{\bar{A}\Delta(u/n)y}{n_k' u_k' \bar{A}_{Stop} y_{Stop}}\right)_i^2}, \qquad (12.7)$$

where $\bar{A}_{Stop}$ is the refraction invariant, and y_{Stop} is the marginal ray height at the stop aperture. These factors make S independent of field of view, lens scaling, and $F/\#$.

The parameters W and S, thus quantify the power distribution and symmetry in a lens system, and lens stress and relaxation. The smaller the parameters are, the less aberration and stress is expected in a lens system. They also provide some information about lens form. Table 12.1 provides W and S for some lens forms.

Large angles of ray incidence lead to increased higher orders of aberration. Then a metric of lens stress/relaxation, R, is based on a real ray surface refraction invariant, $n \sin(I)$, where I is the real angle of incidence on a surface of the marginal, chief or other chosen ray, and is written as,

$$R = \sqrt{\frac{1}{k} \sum_{i=1}^{k} (n \sin(I))_i^2}. \qquad (12.8)$$

Table 12.1 *Typical parameters* W *and* S *for some lens forms*

Lens form	W	S
Cooke triplet	1.12	0.89
Petzval objective (stop at first doublet)	0.63	0.62
Double Gauss	1.01	0.16
Telephoto	0.72	1.30
Reversed telephoto	1.16	0.98
Fisheye	0.23	0.19
Micro-lithography relay	0.28	0.19
Microscope objective	0.23	0.11

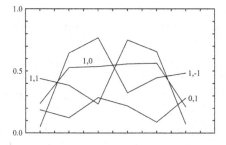

Figure 12.3 Values of $n \sin (I)$ on each surface of a Cooke triplet lens for the marginal ray ($H_y = 0$, $\rho_y = 1$), the chief ray ($H_y = 1$, $\rho_y = 0$), and the rays ($H_y = 1$, $\rho_y = \pm 1$). The end points of each segment, in each line, indicate the values $n \sin (I)$ for each of the six surfaces in a Cooke triplet lens.

Relaxed lenses mitigate higher order aberration by splitting the optical power among several lenses. They tend to have wide tolerances. In contrast, stressed lenses combine strong positive and negative optical power with the benefit of creating degrees of freedom for correcting aberration, such as Petzval field curvature, at the expense of introducing large amounts of higher order aberration, and narrow tolerances.

Figure 12.3 shows values for the surface refraction invariant, $n \sin (I)$, for a Cooke triplet lens. Each line corresponds to a different meridional ray, the marginal ray ($H_y = 0$, $\rho_y = 1$), the chief ray ($H_y = 1$, $\rho_y = 0$), and the rays ($H_y = 1, \rho_y = \pm 1$). Large values of $n \sin (I)$ give place to higher order aberration and, thus, measure lens stress. Surfaces 3 and 4 have the largest values of $n \sin (I)$.

The calculation of the W and S parameters, and the plots of $n \sin (I)$ in Figure 12.3, were done by writing a macro language program that was executed within a lens design program.

Lenses that have degrees of freedom to correct for Petzval field curvature, do not necessarily fully correct for it. A full correction would mean more lens stress, and larger higher order aberrations. By leaving the Petzval sum uncorrected and balancing aberration, an optimum status of correction can be reached. In such lenses, like the Cooke triplet, the ratio of the Petzval radius to the focal length, $\rho_{Petzval}/f$, provides a measure of the lens forms ability to stand lens stress.

12.3 Petzval Portrait Objective

The Petzval portrait objective consists of two separated achromatic doublets. Negative coma is contributed by the rear doublet, and, since the stop is located at the front doublet, enough negative astigmatism can be produced to artificially flatten the field, or to correct for astigmatism. Positive coma aberration in the front doublet corrects for negative coma in the rear doublet, and spherical aberration is individually corrected at each doublet. There are no degrees of freedom to correct for Petzval field curvature. However, by adding a field flattener lens, a flat-field Petzval portrait objective can be obtained, as shown in Figure 12.4, with $f' = 100$ mm, FOV $= \pm 12°$, $F/4$.

The shape of the field flattener lens linearly influences pupil coma, $\bar{W}_{131}$, and, since pupil coma is related to image distortion, $\bar{W}_{131} = W_{311} + \frac{1}{2}\mathcal{K}\cdot\Delta(\bar{u}^2)$, changing the shape of the field flattener lens can correct for image distortion, W_{311}. This correction results in a slightly concave image side surface of the field flattener lens.

The starting lens for the design shown in Figure 12.4 was the prescription of the original Petzval portrait objective, which was scaled to a focal length of 100 mm. The optimization was done by minimizing the RMS spot size for three field points, $0°$, $8°$, and $12°$, by targeting the focal length to 100 mm with a weight of 100, and targeting distortion aberration with a weight of 1. The stop position was initially left at the front doublet, and then it was allowed to vary. Three wavelengths, F, d, and C, were used to optimize the lens. No packaging constraints were set in the merit function, and lens thicknesses were adjusted manually. The front doublet was separated into two singlet

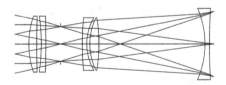

Figure 12.4 Petzval portrait objective with a field flattener lens. $f' = 100$ mm, FOV $= \pm 12°$, $F/4$.

Table 12.2 *Constructional data of the flat field Petzval objective in mm.*
f′ = 100 mm, FOV = ±12°, F/4. W = 0.76, S = 0.94

Surface	Radius	Thickness	Glass
1	54.0796	5.25	N-PSK58
2	529.0112	3	Air
3	−130.0950	3	N-KZFS8
4	−892.5498	9.884	Air
Stop		14.4743	Air
6	160.4580	3	N-KZFS5
7	33.2739	2	Air
8	37.6057	6	N-SK4
9	−104.4887	69.7654	Air
10	−56.7590	3	N-BK7
11	388.3197	3	Air

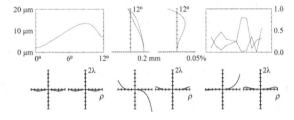

Figure 12.5 Flat field Petzval portrait objective. The RMS spot size vs. field, field curves, and plots of $n \sin(I)$ for upper and lower marginal rays at full field (top). Meridional and sagittal wave fans for the 0°, 8°, and 12°, field points for $\lambda = 587$ nm (bottom).

lenses to allow for independent control of spherical aberration and coma. Glass choice was selected by the optimizer, and is given along the lens constructional data in Table 12.2. Except for the radius of curvature of the second surface of the rear doublet, which is the smallest radius, all other radii were allowed to vary. In addition, the doublet separation and the distance to the image plane were allowed to vary. In the last optimization runs, all the radii of curvature, and the stop position, were allowed to vary.

Figure 12.5 (top) shows the RMS spot size vs. field, field curves, plots of $n \sin(I)$, and (bottom) wave fans for the 0°, 8°, and 12° field points. The wave fans indicate that the design is limited by W_{131}, W_{331}, and W_{333} odd aberrations, and other higher-order odd aberrations of the same type. The plots of $n \sin(I)$ show large values at the intermediate surface of the rear doublet, which coincide with high values of W_{131}, W_{331}, and W_{333} aberrations. A disadvantage of a field flattener lens is its relatively large size and small back focal length. There is no light vignetting in this design.

The Petzval portrait objective is a relaxed lens form, and it is a good choice for solving many imaging problems. Many variations of the Petzval lens are possible by reversing the doublets and lens splitting to increase optical speed.

12.4 Cooke Triplet Lens

The Cooke triplet lens shown in Figure 12.6 was invented by H. D. Taylor, who worked for T. Cooke & Sons of York, as described in US Patents 540,122 (1895) and 568,052 (1896). The essential design goal of the lens design was to obtain a flat image substantially free from astigmatism, distortion, and chromatic aberrations, with the minimum number of lenses possible.

Taylor reasoned that two lens elements of the same but opposite optical power would be free from Petzval field curvature, and that separating them would produce optical power, as showed in Figure 12.7. With the stop at the negative lens and correcting for coma aberration at each lens, the combination is almost free from astigmatism aberration. The astigmatism of the negative lens is nearly corrected with the astigmatism of the positive lens. The positive lens having the stop remote adds astigmatism due to its coma and spherical aberration according to stop shifting,

$$S_{III}^* = S_{III} + 2 \cdot \bar{S} S_{II} + \bar{S}^2 S_I. \tag{12.9}$$

However, as the positive lens is corrected for coma, the additional astigmatism is from spherical aberration. This is not large, as the lens shape for zero coma nearly coincides with the lens shape for minimum spherical aberration.

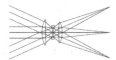

Figure 12.6 Cooke triplet lens. $f' = 50$ mm, FOV $= \pm 24°$, $F/5$.

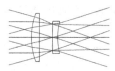

Figure 12.7 A positive and a negative lens of equal but opposite optical power independently corrected for coma aberration. The lens separation provides optical power, and Petzval field curvature is corrected.

By splitting the positive lens into two, and arranging them symmetrically about the negative lens, a flat-field combination free from coma, distortion, and Petzval field curvature results. The spherical aberration and residual astigmatism is then corrected by changing the lens shapes. The Cooke triplet lens was the result of thinking "out-of-the-box." Taylor stated that his lens did not use diaphragm correction, i.e. artificially flattening the field, but that it achieved a flat field with no astigmatism at the edge of the field. The Cooke triplet lens provides enough degrees of freedom to correct for all the primary aberrations. However, in practice, they are not fully corrected because they are balanced against higher order aberrations. Petzval field curvature in the Cooke triplet lens is controlled by creating a bean constriction on the negative lens, and two bulges, one on each of the positive lenses.

Theoretically, to design a Cooke triplet, the following thin lens equations need to be solved for the total power ϕ of the lens, for chromatic change of focus, for chromatic change of magnification, and for Petzval field curvature, respectively,

$$y_a\phi_a + y_b\phi_b + y_c\phi_c = y_a\phi, \tag{12.10}$$

$$\frac{y_a^2}{v_a}\phi_a + \frac{y_b^2}{v_b}\phi_b + \frac{y_c^2}{v_c}\phi_c = 0, \tag{12.11}$$

$$\frac{y_a\bar{y}_a}{v_a}\phi_a + \frac{y_b\bar{y}_b}{v_b}\phi_b + \frac{y_c\bar{y}_c}{v_c}\phi_c = 0, \tag{12.12}$$

$$\frac{\phi_a}{n_a} + \frac{\phi_b}{n_b} + \frac{\phi_c}{n_c} = 0. \tag{12.13}$$

In practice, the design of a Cooke triplet takes advantage of the power of an optimizer in a lens design program. For example, the index of refraction for the lens elements is selected at about $n = 1.6$, as there are many glasses with this index that have v-numbers ranging from about 35 to about 68. This enables us, later in the design, to choose different glasses for correcting chromatic aberration without disrupting the monochromic correction which is done first. Then the positive lenses are set with half the power of the negative lens and arranged symmetrically, and separated from the negative lens to achieve optical power. The lens system is corrected for spherical aberration, coma, and astigmatism using aberration coefficients in the merit function with weights of 1 each, while targeting the focal length to, say, 50 mm with a weight of 100. The surface curvatures are used as variables. This allows the optimizer to find a primary aberration solution. This solution is then optimized for minimum spot size with real rays. The solution is then further modified by incremental steps to meet the specifications for the lens. An optimized solution may not correct exactly for

Petzval field curvature. This is to minimize the optical power of the lens elements, to reduce the lens system optical stress, and to provide less overall aberration.

The chromatic correction is done at the end by changing the glass choice among those glasses with about the same index. Crown glass is used for the positive lenses, and flint glass is used for the negative lens. Since the Cooke triplet is nearly symmetrical, little effort is required to correct for chromatic change of magnification. The chromatic change of focus is corrected by changing the glass v-number difference between the positive and negative lenses. By also allowing the relative air space between the lenses to change, distortion aberration is corrected. There are two solutions to the Cooke triplet according to the shape of the negative lens. One can be obtained from the other by reversing all the elements of one solution and re-optimizing the lens.

A preliminary solution for a Cooke triplet lens can be obtained by following the twelve steps given below. The method relies in the fact that spherical aberration is mainly controlled by the focal length of the negative lens, coma and astigmatism are controlled by the shapes of the positive lenses, and distortion aberration is controlled by the relative airspace between lenses. The chromatic aberrations are controlled by the glass v-number choice, while maintaining about the same nominal index of refraction.

1. Set the entrance pupil diameter to 25 units and the field of view to $\pm 24°$. Start the design monochromatically.
2. Choose a high index glass for the positive elements, such as N-LAK33. Choose a glass for the negative element with about the same index of refraction but different v-number. For example, glass TIF6.
3. Set a positive lens with a focal length of 100 units and a concave-plano lens with a focal length of -100 units. The stop aperture is set at the negative lens. To obtain about half the optical power the airspace between lenses is set to 12 units. Optimize for zero fourth-order astigmatism, W_{222}, using the radii of curvature as variables, but leaving the rear surface of the negative lens as planar.
4. Correct for spherical aberration, W_{040}, by changing the focal length of the negative concave-plano lens, say to -153 units.
5. Add a rear positive lens with N-LAK33 glass with a focal length of 100 units and separated from the negative lens by 12 units. Lens thicknesses are, for example, 5, 2, and 5 units, respectively.
6. Using the rear curvature of the negative lens and the curvatures of the third positive lens, correct for astigmatism, W_{222}, aberration, and for a focal length of the negative lens half as large, i.e. -76.5 units.

7. Maintaining the focal length of the lenses, and using only the shapes of the positive lenses, correct for astigmatism, W_{222}, and coma, W_{131}, aberration.

8. While maintaining the correction for coma and astigmatism, correct for spherical aberration, W_{040}, by changing the focal length of the positive lenses from 100 units to, say, 120 units. The shape of the negative lens is not changed.

9. While maintaining the correction for spherical aberration coma and astigmatism, correct for distortion aberration, W_{311}, by changing the second airspace to, say, 13.5 units. The shape of the negative lens, and the lens focal lengths are not changed.

10. While maintaining zero coma, and by changing only the shapes of the positive lenses, introduce negative, or positive, astigmatism to flatten the tangential field curve.

11. Add the desired wavelengths and change the glass of the negative element to correct for chromatic change of focus, while maintaining the same nominal index of refraction. Change the glass of one of the positive lens elements to correct any chromatic change of magnification residual, while maintaining the same nominal index of refraction.

12. Scale the lens system to the desired focal length, change the relative aperture and field of view as desired. The lens is ready for real ray optimization.

Alternatively, a Cooke triplet can be designed starting from parallel plates, or with a global search optimizer, or starting from a lens in the patent literature. However, a lens designer should know how to design a lens from first principles.

The prescription for the lens in Figure 12.6 is given in Table 12.3, and the performance curves are shown in Figure 12.8. In this lens, to reduce aberration residuals, higher index glass, N-LAK33, was used for the positive elements.

Table 12.3 *Constructional data of the Cooke triplet lens,*
$f' = 50$ *mm, FOV* $= \pm 24°$, F/5

Surface	Radius (mm)	Thickness (mm)	Glass
1	30.1516	3.25	N-LAK33
2	−677.4312	6.0	Air
3 (Stop)	−23.3371	1.0	TIF6
4	29.3810	4.75	Air
5	5004.4670	3.0	N-LAK33
6	−20.7956	43.1266	Air

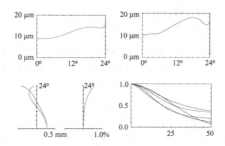

Figure 12.8 Cooke triplet lens. Top left, RMS spot size vs. field of field. Top right, radius that encircles 80% of the energy vs. field of view. Bottom left, field curves. Bottom right, MTF curves in lines/mm for fields 0°, 17°, and 24°. $f' = 50$ mm, FOV = ±24°, F/5.

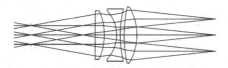

Figure 12.9 An external stop and telecentric Cooke triplet lens. $f' = 50$ mm, FOV = ±10°, F/5.

The ratio of the Petzval radius to the focal length is, $\rho_{Petzval}/f' = -3.2$, $W = 0.99$, and $S = 0.74$. Keeping track of the lens elements focal length helps to understand the structure of the lens system. For the Cooke triplet in Table 12.3, the focal lengths are $f_{1'} = 38.2$ mm, $f_{2'} = -20.8$ mm, and $f_{3'} = 27.36$ mm. Thus, the lens elements have more optical power than the lens system.

The lens was first optimized using RMS spot size for image quality. However, this metric is optimistic because the radius that encloses 80% of the energy is larger than the RMS spot radius for the same field point. The astigmatism field curves show the typical crossing indicating a good balance of orders of astigmatism, field curvature, and focus. Further optimization to approach meeting the 50/30 MTF rule of thumb was done by optimizing with RMS wavefront error and targeting MTF values at 30 lines per millimeter. The F, d, and C wavelengths were used for optimization and analysis.

As shown in Figure 12.9, the Cooke triplet lens can be modified so that it is telecentric in image space. The field of view is reduced and the stop aperture is moved to the front. This is done in several small steps, in which the lens is re-optimized at each step to keep the local minimum within the reach of the optimizer. The stop aperture is external to the lens, and this allows for combining two such lenses to form an optical relay.

12.5 Double Gauss Lens

Paul Rudolph, from the Zeiss Company in Germany, invented the planar lens, as shown in Figure 12.10, US Patent 583,336, (1897). It is a symmetrical lens about the stop aperture. Rudolph used the Gauss objective as a starting point for designing the rear part of his planar lens. In order to control Petzval field curvature, he thickened the negative meniscus lens of the Gauss objective, and used the shape of the lenses and distance to the stop aperture to correct for the even aberrations, spherical aberration, astigmatism, and field curvature. To mitigate the odd aberrations, coma, and distortion, he used the technique of doubling a lens about the stop aperture, making the lens symmetrical.

Finally, Rudolph came up with a novel technique to correct for the chromatic aberrations. He used a buried surface, one that has the same index of refraction on both sides but different v-numbers, on two lenses to create two dispersive interfaces and to control chromatic change of focus and magnification. The result was a lens capable of a high speed and a wide field of view, with excellent imaging. The name double Gauss lens refers to lenses that have been derived from the planar lens by splitting lens elements and by breaking the symmetry about the stop aperture. At the time of the invention there were no antireflection coatings, and the eight air-to-glass interfaces caused an increased amount of light loss by Fresnel reflection, and a decreased image contrast due to the Fresnel reflections reaching the image plane. Double Gauss type lenses have been used widely for 35 mm photographic cameras.

Figure 12.11 shows a double Gauss lens design described as Example 1 in US Patent 4,123,144 (1978). Of note is the significant light vignetting that takes place which is used to help control oblique spherical aberration, W_{240}, and packaging constraints. One constraint in lenses for 35 mm Single Lens Reflex cameras is the back focal length, which must be large enough to allow for the folding mirror that is required. This constraint imposes a limit on how well the aberrations can be controlled. Figure 12.12 shows the wave fans for the $0°$, $17°$, and $22°$ field positions. The top row shows the wave fans for full speed at $F/2$ with a scale of 10λ at 587 nm, and the bottom row for a speed of $F/5.6$, with a scale of 2.5λ. There is a significant amount of oblique spherical

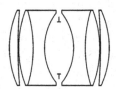

Figure 12.10 Planar lens from Paul Rudolph. US Patent 583,336, example 2.

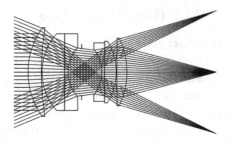

Figure 12.11 Double Gauss lens, US Patent 4,123,144, example 1. $f' = 50$ mm, FOV = ±22°, F/2. Note the light vignetting which helps to clip aberrated rays and to decrease the size of the rear lens elements.

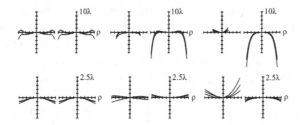

Figure 12.12 Meridional and sagittal wave fans for double Gauss lens, US Patent 4,123,144, example 1. Top row, at full F/2 speed. Bottom row, at F/5.6 speed. Field positions are 0°, 17°, and 22°.

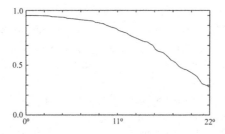

Figure 12.13 Relative illumination for the double Gauss lens, US Patent 4,123,144, example 1.

aberration at full speed, which in part is controlled by light vignetting. This does not mean that the lens is not well designed. On the contrary, given the packaging constraints and an understanding of the application, the designer provided excellent imaging at F/5.6, knowing that a photographer would first focus the lens at F/2 and then switch to F/5.6 for best image quality. Oblique spherical aberration is a limiting aberration in double Gauss type lenses. Figure 12.13 shows the relative illumination at F/2, which decreases to 0.3

Table 12.4 *Prescription of the Double Gauss lens, US Patent 4,123,144, example 1.* f' = 100 mm, FOV = ±22°, F/2

Surface	Radius	Thickness	Glass
1	48.88	8.89	1.62286/60.08
2	182.96	0.38	Air
3	36.92	15.11	1.58565/46.17
4	∞	2.31	1.67764/31.97
5	23.06	9.14	Air
Stop		13.36	Air
7	−23.91	1.92	1.57046/42.56
8	∞	7.77	1.64128/55.15
9	−36.92	0.38	Air
10	1063.24	6.73	1.62286/60.08
11	−43.88	59.18	Air

at the edge of the field. Usually for photography the relative illumination is specified at full field to be 0.5 (50%) or larger.

The focal lengths of the external positive lenses and of the negative menisci lenses are $f_{1'} = 52.02$ mm, $f_{2'} = -58.73$ mm, $f_{3'} = -123.32$ mm, and $f_{4'} = 37.47$ mm, and the focal length of the lens system is $f' = 50$ mm. The parameters are $W = 1.02$ and $S = 0.2$, and the Petzval radius is -275.8 mm.

The prescription of the lens is given in Table 12.4. The buried surfaces were left as plano surfaces. However, the radius of curvature of a buried surface is an effective degree of freedom to control chromatic aberration. Alternatively, the v-number difference of the glasses surrounding the buried surface is also an effective degree of freedom.

Further Reading

Hopkins, R. E. "Third-order, and fifth-order analysis of the triplet," *Journal of the Optical Society of America*, 52(4) (1962), 389–94.

Isshiki, Masaki, Sinclair, Douglas C., Kaneko, Seiichi. "Lens design: global optimization of both performance and tolerance sensitivity," Proceedings of SPIE 6342, International Optical Design Conference 2006, 63420N (2006), doi: 10.1117/12.692252.

Jonas, Reginald P., Thorpe, Michael D. "Double Gauss lens design: a review of some classics," Proceedings of SPIE 6342, International Optical Design Conference 2006, 634202 (2006), doi: 10.1117/12.692187.

Mandler, Walter. "Design of basic double gauss lenses," Proceedings of SPIE 0237, 1980 International Lens Design Conference (1980), doi: 10.1117/12.959089.

Mandler, W., Edwards, G., Wagner, E. Four-member Gauss objective, US Patent 4,123,144 (1978).

Rudolph, P. Objective glass, US Patent 583,336 (1897).

Sasián, J. M., Descour, M. R. "Power distribution and symmetry in lens systems," *Optical Engineering*, 37(3) (1998), 1001–4.

Shafer, David. "Optical design and the relaxation response," Proceedings of SPIE 0766, Recent Trends in Optical Systems Design and Computer Lens Design Workshop (1987), doi: 10.1117/12.940196.

Sharma, K. D., Rama Gopal, S. V. "Significance of selection of Petzval curvature in triplet design," *Applied Optics*, 21 (1982), 4439–42.

Taylor, H. D. Lens, US Patent 540,122 (1895)
    Lens, US Patent 568,052 (1896).

13

Lens System Combinations

An optical engineer is not only concerned with the design of a single lens system, but also in combining several lens systems. An optical system may comprise several individual lens systems. These lens systems must be combined to meet the overall optical system specifications. Often each lens system serves to relay an image or a pupil of the previous system to a new location. In combining lens systems, several effects can take place due to image and/or pupil aberrations. Being aware of such effects is key to design, analyze, or debug a combination of lens systems. For example, a telescope can be considered as the combination of an objective lens, an image erecting system, an eyepiece, and the human eye. To properly form an image on the eye's retina, these subsystems must be properly combined. In this chapter we discuss combining lens systems, pupil aberrations, and optical relays.

13.1 Image Aberrations

We assume that, in combining two lens systems, we maintain axial symmetry and that both systems work at their designed optical conjugates. This situation can be described as first-order image matching. If the image aberration functions of those two systems, A and B, are $W_A(\vec{H}, \vec{\rho})$ and $W_B(\vec{H}, \vec{\rho})$, then the aberration function of the lens combination, $W_C(\vec{H}, \vec{\rho})$, is,

$$W_C\left(\vec{H}, \vec{\rho}\right) = W_A\left(\vec{H}, \vec{\rho}\right) + W_B\left(\vec{H}, \vec{\rho}\right) + W_{AB}\left(\vec{H}, \vec{\rho}\right), \qquad (13.1)$$

where $W_{AB}(\vec{H}, \vec{\rho})$ are called extrinsic, or induced, aberrations. These aberrations result from the synergy of the combination of the systems. If in the aberration functions of systems, A and B, there are fourth-order aberration terms, then the extrinsic aberrations are of sixth-order.

153

Table 13.1 *Pupil aberration and image aberration coefficient connections*

$\bar{W}_{040} = W_{400}$	$\bar{W}_{220} = W_{220} + \dfrac{1}{4}\mathcal{K}\cdot\Delta\{u\bar{u}\}$
$\bar{W}_{131} = W_{311} + \dfrac{1}{2}\mathcal{K}\cdot\Delta\{\bar{u}^2\}$	$\bar{W}_{311} = W_{131} + \dfrac{1}{2}\mathcal{K}\cdot\Delta\{u^2\}$
$\bar{W}_{222} = W_{222} + \dfrac{1}{2}\mathcal{K}\cdot\Delta\{u\bar{u}\}$	$\bar{W}_{400} = W_{040}$

In addition, if between systems A and B there is a displacement in the field, $\vec{H}_0$, or in the aperture, $\vec{\rho}_0$, then the axial symmetry is broken, and the aberrations of the combination can be written, neglecting extrinsic terms, as,

$$W_C\left(\vec{H},\vec{\rho}\right) = W_A\left(\vec{H},\vec{\rho}\right) + W_B\left(\vec{H} + \vec{H}_0, \vec{\rho} + \vec{\rho}_0\right). \qquad (13.2)$$

For simplicity, we will assume that first-order image matching takes place.

The aberrations of a lens system can be divided according to the algebraic power of the aperture into even and odd aberrations. The odd aberrations tend to cancel whenever there is some symmetry about the stop aperture in a lens system. However, the even aberrations, or even aberration residuals, may add and accumulate when combining several lens systems. In particular, secondary chromatic change of focus is an even aberration, and any residual will likely add. Thus, in combining several lens systems, one must minimize and be aware of the image aberration residuals of the constituting lens systems.

13.2 Pupil Aberrations

In the same way the object and the image are optically conjugated, the entrance and the exit pupil are also conjugated. Therefore, the pupils, one being the image of the other, can suffer from aberration. The pupil aberration function, $\bar{W}(\vec{H},\vec{\rho})$, to fourth order is,

$$\bar{W}\left(\vec{H},\vec{\rho}\right) = \bar{W}_{000} + \bar{W}_{200}\left(\vec{\rho}\cdot\vec{\rho}\right) + \bar{W}_{111}\left(\vec{H}\cdot\vec{\rho}\right) + \bar{W}_{020}\left(\vec{H}\cdot\vec{H}\right)$$
$$+ \bar{W}_{040}\left(\vec{H}\cdot\vec{H}\right)^2 + \bar{W}_{131}\left(\vec{H}\cdot\vec{H}\right)\left(\vec{H}\cdot\vec{\rho}\right) + \bar{W}_{222}\left(\vec{H}\cdot\vec{\rho}\right)^2$$
$$+ \bar{W}_{220}\left(\vec{H}\cdot\vec{H}\right)\left(\vec{\rho}\cdot\vec{\rho}\right) + \bar{W}_{311}\left(\vec{\rho}\cdot\vec{\rho}\right)\left(\vec{H}\cdot\vec{\rho}\right)$$
$$+ \bar{W}_{400}\left(\vec{\rho}\cdot\vec{\rho}\right)^2. \qquad (13.3)$$

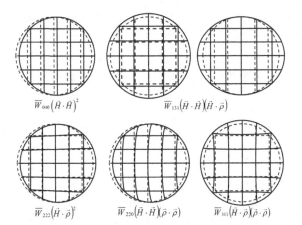

$$\overline{W}_{040}\left(\vec{H}\cdot\vec{H}\right)^2 \qquad \overline{W}_{131}\left(\vec{H}\cdot\vec{H}\right)\left(\vec{H}\cdot\vec{\rho}\right)$$

$$\overline{W}_{222}\left(\vec{H}\cdot\vec{\rho}\right)^2 \qquad \overline{W}_{220}\left(\vec{H}\cdot\vec{H}\right)\left(\vec{\rho}\cdot\vec{\rho}\right) \qquad \overline{W}_{311}\left(\vec{H}\cdot\vec{\rho}\right)\left(\vec{\rho}\cdot\vec{\rho}\right)$$

Figure 13.1 Pupil grid mapping effects due to pupil aberrations in relation to the ideal first-order pupil (dashed line grid). There is no effect from pupil piston aberration.

The primary pupil aberrations of concern are spherical aberration, coma, astigmatism, field curvature, and distortion. The pupil aberration coefficients are connected to the image aberration coefficients, as shown in Table 13.1.

Pupil aberrations are interpreted as a distortion of the cross-section of the light beam from a field point, when the cross-section of the beam at the other pupil is undistorted. Figure 13.1 shows the effects of pupil aberrations on the cross-section of a beam at a pupil. The dashed line grid represents the ideal undistorted first-order image of the pupil.

If two lens systems, A and B, are combined, the exit pupil of the first must be coincident with the entrance pupil of the second. There are six first-order degrees of freedom in connecting two systems; three translations, X, Y, and Z, and three rotations, α, β, and γ. If the lens system combination preserves axial symmetry, and if the location of the exit pupil of system A coincides with the location of the entrance pupil of system B, we have pupil matching to first-order. Any departure from this situation can cause light loss by vignetting, or reduce the field of view, or cause image aberration. For simplicity, we assume there is pupil matching to first-order.

If system A suffers from image primary aberrations, $W_A(\vec{H}, \vec{\rho})$, and system B suffers from pupil primary aberrations, $\bar{W}_B(\vec{H}, \vec{\rho})$, then the extrinsic aberrations, $W_{AB}(\vec{H}, \vec{\rho})$, of sixth-order are given by,

$$W_{AB}\left(\vec{H}, \vec{\rho}\right) = -\frac{1}{\mathcal{K}}\vec{\nabla}_{\rho}W_A\left(\vec{H}, \vec{\rho}\right)\cdot\vec{\nabla}_H\bar{W}_B\left(\vec{H}, \vec{\rho}\right), \qquad (13.4)$$

where $\vec{\nabla}_{\rho}W_A(\vec{H}, \vec{\rho})$ is the gradient of the image aberration function and $\vec{\nabla}_H\bar{W}_B(\vec{H}, \vec{\rho})$ is the gradient of the pupil aberration function. Despite there being pupil matching to first-order, the presence of image aberrations in system A and pupil aberrations in system B produces extrinsic aberration. For example, if system A has chromatic aberrations, $\partial_\lambda W_{020}(\vec{\rho} \cdot \vec{\rho})$ and $\partial_\lambda W_{111}(\vec{H} \cdot \vec{\rho})$, and system B has pupil distortion, $\bar{W}_{311}(\vec{H} \cdot \vec{\rho})(\vec{\rho} \cdot \vec{\rho})$, then the extrinsic terms are,

$$
W_{AB}\left(\vec{H}, \vec{\rho}\right) = -\frac{1}{\mathcal{K}}\left(\partial_\lambda W_{020}\bar{W}_{311}\left(\vec{\rho} \cdot \vec{\rho}\right)^2 + \partial_\lambda W_{111}\bar{W}_{311}\left(\vec{H} \cdot \vec{\rho}\right)\left(\vec{\rho} \cdot \vec{\rho}\right)\right).
$$

(13.5)

The first term, $\partial_\lambda W_{020}\bar{W}_{311}(\vec{\rho} \cdot \vec{\rho})^2$, is spherochromatism, and the second term, $\partial_\lambda W_{111}\bar{W}_{311}(\vec{H} \cdot \vec{\rho})(\vec{\rho} \cdot \vec{\rho})$, is variation of coma with the wavelength, which can be called comachromatism. Thus, in combining lens systems, it is important to have them individually corrected for the chromatic aberrations. Table 6 in Appendix 1 provides extrinsic aberrations in the presence of the primary monochromatic aberrations.

13.3 Pupil Spherical Aberration

Pupil spherical aberration, $\bar{W}_{040}$, causes the beam from a given field point at a pupil to translate perpendicularly to the optical axis. This effect depends on the cube of the field of view. In combining two lens systems under the presence of pupil spherical aberration, some light vignetting can take place. Figure 13.2 (top) shows light beams from different field points converging to the exit pupil of a lens, but suffering from pupil spherical aberration. These beams might pass through the entrance pupil of the subsequent lens system and illuminate

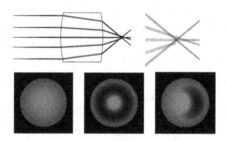

Figure 13.2 Top row, light beams converging to a lens exit pupil suffer from pupil spherical aberration; ray detail (right). The beams are vignetted differently according to the position and size of the entrance pupil of a lens system. Bottom row, obscuration of the field of view. Left, no obscuration. Center, annular obscuration (caused by axial displacement). Right, kidney bean obscuration (caused by lateral displacement).

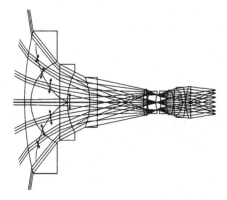

Figure 13.3 A fisheye lens system covering a field of view of ±90° at *F*/2. The entrance pupil for each field point changes size and position along the external caustic sheet for spherical aberration, as shown by the double arrowhead lines on the caustic line.

completely the field of view. However, if the entrance pupil of the subsequent lens system is moved along the optical axis, light from some field points may miss passing through the entrance pupil, and the field of view would be obscured annularly, as also shown in Figure 13.2. If, in addition, the entrance pupil of the following system is laterally moved, the field of view would be obscured in the form of a kidney bean, as also shown in Figure 13.2. This kidney bean effect is often observed in visual optical systems that provide a large apparent field of view, that suffer from pupil spherical aberration, and when the observer shifts viewing location.

Wide angle lenses, such as fisheye lenses, suffer from spherical aberration of the entrance pupil, as illustrated in Figure 13.3. The entrance pupil for different field points changes size and position. The phenomenon is known as pupil walking, and it enables a lens to accept light for fields of view beyond 90°. While pupil walking is an extreme effect in fisheye lenses, it draws attention to matching pupils when combining a wide angle lens system.

Figure 13.4 shows a lens with the stop in front and five principal rays. The stop aperture is located at the front focal point of the lens and, to first-order, the lens is telecentric in image space. Because of pupil spherical aberration, the slope of the rays in image space changes as a function of the lens aperture. This change of slope represents loss of telecentricity, and can result in light loss or aberrations when combining lens systems.

In photographic lenses, pupil spherical aberration can help to reduce the angle of incidence of the chief ray in the image plane to meet lens specifications. For example, lenses for mobile phones, as shown in Figure 13.5, use

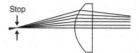

Figure 13.4 A lens with the stop aperture located at the front focal point is to first-order telecentric in image space, and principal rays should be parallel to the optical axis after they refract through the lens. Due to spherical aberration of the pupil, the principal rays, as a function of the aperture, progressively acquire a non-zero slope with respect to the optical axis.

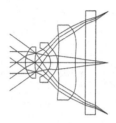

Figure 13.5 Three lens element and infrared filter lens system for a mobile phone. The chief ray decreases in slope after being refracted by the aspheric third lens element.

aspheric surfaces. The position of the stop near the front of the lens, and the last strongly aspheric lens element, which turns from negative to positive power at the lens edge, helps to reduce the chief ray angle of incidence in the image plane. An aspheric surface near the image plane is an effective degree of freedom to control pupil spherical aberration.

13.4 Pupil Coma

Pupil coma, $\bar{W}_{131}$, causes the cross-section of a beam at a pupil to change size and to be anamorphically distorted. Figure 13.6 shows two afocal optical relays using ideal lenses and a plano-convex field lens to control the path of rays. The difference between the two relays is the shape of the plano-convex lens, $X = -1$ or $X = 1$, and the slightly different path for the rays for the off-axis field point. Pupil coma depends linearly on the lens shape, X, and in this case it is positive for $X = -1$, and negative for $X = 1$. For reference, the footprint of the on-axis beam is shown in Figure 13.6 (bottom-center). If pupil coma is positive, the cross-section of the off-axis beam expands, as shown in Figure 13.6 (bottom-right), and if it is negative it contracts, as also shown in Figure 13.6 (bottom-left). In addition, the field lens contributes pupil spherical aberration; this aberration makes the real chief ray cross the optical axis ahead of the second ideal lens, and displaces the beam laterally.

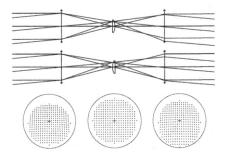

Figure 13.6 Two afocal optical relays with a plano-convex field lens and two ideal lenses. The stop aperture is at the first ideal lens. Light rays are shown for on-axis and off-axis field points. Off-axis beam footprints at the second ideal lens are shown at the bottom.

Figure 13.7 When the stop aperture coincides with the entrance pupil of a lens, or is buried in the lens, the exit pupil can change size and be anamorphically distorted due to pupil coma. As shown, from left to right, this effect increases with the square of the field of view.

The fact that the footprint for off-axis beams expands or contracts can result in light vignetting and image aberration, when light is directed to the entrance pupil of a lens system that follows the lens system suffering from pupil coma. The relative illumination, $RI(\vec{H})$, of a lens also depends on pupil coma,

$$RI\left(\vec{H}\right) = 1 - \left(2\bar{u}'^2 - \frac{4}{\not{K}}\bar{W}_{131}\right)\left(\vec{H}\cdot\vec{H}\right). \qquad (13.6)$$

As a result of pupil coma, off-axis beams change size and are anamorphically distorted, as shown in Figure 13.7. This beam expansion or contraction changes the convergence of the beam at the focal plane of a lens system, and, therefore, also changes the theoretical resolution and MTF cut-off spatial frequency. Figure 13.8 shows a singlet lens with a strong amount of pupil coma. The off-axis beam shows a significant increase in its optical speed in comparison to the on-axis beam.

13.5 Pupil Distortion

Pupil distortion, $\bar{W}_{311}$, causes the image of the pupil to be barrel or pincushion distorted, as shown in Figure 13.9. Because the image of the pupil can be

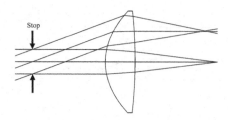

Figure 13.8 Singlet lens with a strong amount of pupil coma. The off-axis beam focuses with a higher optical speed than the on-axis beam.

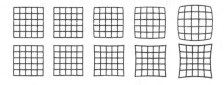

Figure 13.9 Distortion aberration represented by imaging a square grid. Top row, barrel distortion, bottom row pincushion distortion, ranging from left to right, 0%, 1%, 3%, 5%, and 10%.

larger than the entrance pupil of a system, light loss by ray vignetting may take place.

In a lens system, pupil distortion and image coma are related by,

$$\bar{W}_{311} = W_{131} + \frac{1}{2}\mathcal{K}\cdot\Delta\{u^2\}. \tag{13.7}$$

Therefore, in a lens system that is corrected for image coma aberration, there may necessarily be some amount of pupil distortion whenever $u'^2 - u^2 \neq 0$. In addition, since the exit pupil may become larger or smaller than the first-order pupil size, the optical speed of the lens may slightly change from that predicted by first-order optics.

13.6 Chromatic Vignetting

Some lens systems require an intermediate image where a field stop must be located. In the case of visual instruments, the intermediate image must be corrected for chromatic change of magnification to avoid chromatic vignetting. In this effect, light from field points at the edge of the field is vignetted for some wavelengths and not vignetted for others. Then, when light is relayed to form a final image, the edge of the field can appear bluish, yellowish, or reddish due to the chromatic vignetting. This is an undesirable effect in a

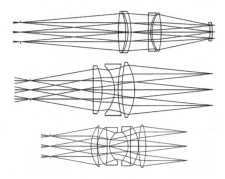

Figure 13.10 External stop and telecentric lenses, $f' = 100$ mm at $F/4$: Petzval objective FOV = $\pm 5°$, Cooke triplet lens FOV = $\pm 10°$, and double Gauss lens FOV = $\pm 10°$. Note that the double Gauss lens is the shortest one in total length.

visual instrument. The presence of other aberrations in an intermediate image where there is a field stop can also produce undesirable effects.

13.7 Optical Relays

Lens optical relays transfer an image, or a pupil, from one location to another location. At the same time lens relays can provide a change of magnification, or satisfy some first-order lens specifications like telecentricity. Lens relays are important in optical engineering, and familiarly with their design and properties should be gained by an optical engineer.

A starting point for the design of an optical relay can be the design of some external stop, telecentric lenses, as shown in Figure 13.10.

Because the lenses shown in Figure 13.10 have their stop aperture external to the lens, and they are telecentric in image space, they can be combined to form afocal relay systems to relay and image or a pupil, as shown in Figure 13.11. By scaling in size one relay, the overall system afocal magnification can be set. Lenses with the stop external to the lens are useful in optical engineering, and creating a collection of such lenses becomes useful over time. Lenses with an external stop allow placement of beam splitters to create two imaging channels.

Some problems in combing lenses to form optical relays are that the aberration residuals may add to degrade the final image. A re-optimization of the complete relay system can restore, or improve, the image quality at the expense of losing lens modularity. Pupil spherical aberration, and pupil coma may create problems in matching the exit and entrance pupils of the lenses that

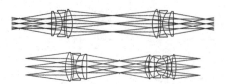

Figure 13.11 Top, pupil relay using Cooke triplet lenses. Bottom, image relay using a Cooke triplet lens and a double Gauss lens. Both relays are afocal and use lenses with $f' = 100$ mm at $F/4$.

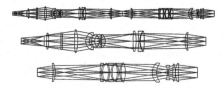

Figure 13.12 Top, double relay system with field lens. There are three images that are represented by vertical straight lines. The first vertical line on the left is an ideal lens forming the first image. The second image is at the center of the relay, and the third image is at the end of the relay at the right. Middle, the first relay; and, bottom, the second relay; these two relays form the complete double relay shown at the top.

are combined. For double telecentric relays we have, $\bar{u}' = \bar{u} = 0$, and by way of equations in Table 13.1 we also have the identities, $\bar{W}_{131} = W_{311}$, $\bar{W}_{222} = W_{222}$, and $\bar{W}_{220} = W_{220}$.

Sometimes it is necessary to add a field lens to control the path of rays so that there is no light vignetting. The field lens may introduce field curvature aberration that will degrade the image quality. A Cooke triplet lens can be designed as a field lens to maintain a flat field. Or the field curvature of the field lens can be corrected elsewhere, at the expense of adding optical stress to the lens system.

Figure 13.12 (top) shows an imaging double relay system working at $F/4$. The first element on the left is an ideal lens that forms a first image. Then there is a field lens made out of high index glass to minimize its Petzval field curvature, which is followed by a plano-parallel glass plate that represents an optical filter. Two afocal relays, shown in Figure 13.12 (middle and bottom), are set one after the other to relay the first image twice. Secondary spectrum aberration from each relay normally adds rather than cancels, but in this design a group of lenses with anomalous dispersion in the second relay helps to mitigate this aberration. Fluorite (FPL53) type and short flint (KZFS1) glasses were used in this group of lenses, which have strong optical power. Both relays were designed as independent well corrected lens subsystems for modularity, alignment, and debugging reasons.

The first attempt to design this relay was using four well corrected and telecentric objective lenses. However, the number of lens elements was excessive and then the relay was simplified by reducing the number of lenses and designed as the combination of two well corrected relays. The menisci thick negative lenses in the first relay help to correct for Petzval field curvature, including that of the first field lens. There are two other field lenses next to the intermediate image. These lenses make the relays telecentric in the space of the center image.

Further Reading

Fallah, Hamid R., Maxwell, Jonathan. "Higher-order pupil aberrations in wide-angle and panoramic optical systems," Proceedings of SPIE 2774, Design and Engineering of Optical Systems (1996); doi: 10.1117/12.246677.

Hoogland, J. Flat field lenses, US Patent 4,575,195 (1986).

Hopkins, H. H. Optical systems, US Patent 4,168,882 (1979).

Lerner, S. A., Kelly, C. D. Optical relay, US Patent 7,175,289 (2007).

Sasián, J. *Introduction to Aberrations in Optical Imaging Systems* (Cambridge, UK: Cambridge University Press, 2013).

Tesar, J. Highly corrected relay system, US Patent 9,918,619 (2018).

Wakimoto, Z., Hayashi, T. Telecentric, image-forming optical system for large image size, US Patent 4,929,066 (1990).

Wetherell, W. B. "Afocal lenses," in *Applied Optics and Optical Engineering*, Vol X, (New York: Academic Press, 1992).

14

Ghost Image Analysis

Lens systems are designed to form an image according to an ideal model. Light that passes through the stop aperture forms the image. However, some light may not contribute to the formation of the intended image, and reaches the image plane to degrade the image. This light is known as stray light, flare, veiling glare, and ghost images.

Ghost images are formed when light is reflected from one or more optical surfaces in a lens system. The number of light reflections, or light bounces, determines the relative amount of optical flux in a ghost image. The light flux depends on the surface reflectivity and decreases roughly in algebraic power as the number of light bounces increases. The light reflection is due to Fresnel reflections or total internal reflection. There are odd ghosts created by an odd number of light reflections that travel back to the object, and even ghosts created by an even number of reflections that travel toward the nominal image plane. A ghost can be an image of the object, an image of the aperture stop, i.e. a pupil, or an image of the light sensor. Veiling glare is contributed by ghost images that are not in focus with the nominal image, and it is often non-uniformly distributed on the image plane.

A singlet lens can have a single two-bounce ghost, a two lens system can have six two-bounce ghosts, a three lens system can have 15 two-bounce ghosts, and the number of two-bounce ghosts increases as $k(k-1)/2$, where k is the number of optical surfaces. Figure 14.1 shows a double convex lens forming an on-axis image of an object at infinity, a single bounce odd ghost, and a two bounce even ghost. When the ghosts images are near the nominal image plane they can result in significant image artifacts. Ghost images usually suffer from high amounts of optical aberration. We are mainly concerned here with two-bounce ghost images. This chapter presents a basic discussion about ghost image analysis.

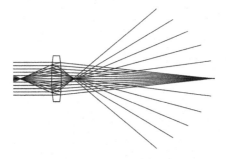

Figure 14.1 A double convex lens forming an image of an object at infinity and two ghost images that are not close to the nominal image plane.

14.1 Surface Reflectivity

At normal incidence and for an air-to-glass interface, the surface reflectance, R, is,

$$R = \left(\frac{n-1}{n+1}\right)^2. \tag{14.1}$$

If the glass index of refraction is $n = 1.5$, then the reflectance is 0.04 or 4%. For a Cooke triplet lens that has three singlet lenses and six air-to-glass interfaces, then the amount of light reflected is roughly 24%. This is a substantial amount of light. Since there are 15 two-bounce reflections, then the amount of light traveling toward the image plane from these reflections is roughly $0.04^2 \times 15 = 0.024$ or 2.4%. This amount of light contributes to veiling glare and ghost images. Veiling glare reduces the image contrast.

To reduce the amount of light reflected by a surface, optical thin film coatings are deposited on the optical surfaces of a lens. Depending on the wavelength range, the angle of incidence, the state of light polarization, and the thin film coating, the surface reflectivity can be decreased. There are single layer and multi-layer thin film coatings that provide different reflectivity, as shown in Figure 14.2. In practice, a single layer thin film coating can reduce the reflectivity to about 1%. Then the amount of light in a double bounce reflection is 0.01%. This amount can be negligible; however, if a ghost image of a bright object such as the sun focuses at the nominal image plane, the ghost can be bright and objectionable. For lens systems used in high power applications, a detailed ghost image analysis must be done to find ghost images that focus in the lenses.

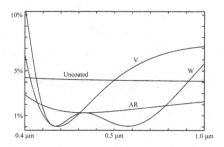

Figure 14.2 Reflectivity at normal incidence of BK7 glass for different anti-reflection coatings: Uncoated, V-coating, W-coating, and single layer AR-coating. The wavelength ranges from 0.4 μm to 1.0 μm.

14.2 First-Order Analysis

Due to the large number of ghost images a first-order analysis is done first. This analysis traces first-order marginal and chief rays, forward and backwards, to determine the ghost image location, its $F/\#$, and the ray height at the nominal image plane for both object and aperture stop ghosts. If a ghost image is deemed to be close to the nominal image plane, then further analysis is done by real ray tracing. Table 14.1 presents a first-order analysis for the Cooke triplet lens of Chapter 12.

On the first column the reflection surfaces are listed, including the image plane as a reflection surface. The closest ghost pupil to the nominal image plane is due to reflection on surfaces seven then five and at a distance of -27.9 mm. The closest ghost image is also due to surfaces seven then five and is at a distance of -23.4 mm. Note the relatively low $F/\#$ for most of the light beams.

14.3 Real Ray Tracing Analysis

Real ray tracing analysis can be done with sequential and non-sequential ray tracing. Here we used sequential ray tracing, and this required preparing a lens file so that the forward ray tracing proceeds until the first Fresnel reflecting surface is reached, then reverse ray tracing proceeds until the second Fresnel reflecting surface is reached, and then forward ray tracing proceeds until the image plane is reached.

Figure 14.3 shows sequential real ray tracing of ghost images from reflection on the image plane, and from surface five of a Cooke triplet lens. The top layout is for the on-axis light beam, and the bottom layout is for an off-axis light beam at 2 degrees.

Table 14.1 *First-order ghost analysis for a Cooke triplet lens*

Reflection surfaces	Distance to ghost pupil, mm	Distance to ghost, mm	F/#
2-1	−53.6	−54.7	0.98
3-2	−73.8	−58.9	1.42
3-1	−276.9	−65.63	1.32
4-3	−54.5	−61.1	0.75
4-2	−75.8	−87.5	1.23
4-1	160.1	−34.7	1.43
5-4	49.9	−34.9	1.4
5-3	−60.6	−64.8	0.98
5-2	−86.1	−101.4	2.26
5-1	−32.6	−43.6	1.25
6-5	−28.3	−40.4	0.66
6-4	−40.1	−51.6	0.97
6-3	−70.5	−53.7	0.91
6-2	−180.2	−74.9	0.95
6-1	−49.8	1556.9	103.8
7-6	−52.5	−51.4	0.97
7-5	−27.9	−23.4	2.4
7-4	−34.9	−38.4	1.28
7-3	−70.1	−67.6	1.43
7-2	−166.7	−122.4	4.7
7-1	−49.5	−47.2	1.45

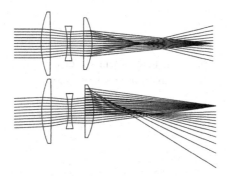

Figure 14.3 Ghost images for an on-axis beam and an off-axis beam at 2° caused by reflection on the image plane and on surface five of a Cooke triplet lens. Note the strong amount of aberration in the ghost images. The primary beam to the image plane is also shown.

Since the light from the ghost is substantially out of focus from the nominal image plane, it will not create an image artifact. However, the light distributed over a relatively large area at the image plane will contribute to veiling glare. If a ghost image is in focus with the nominal image, it might be possible to

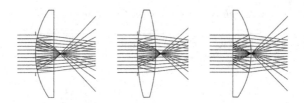

Figure 14.4 For a thin lens the ghost image position due to a two-bounce reflection is nearly independent of lens shape. The focal length of the ghost layout is about seven times smaller than that of the nominal lens.

change the curvatures of the optical surfaces responsible for the ghost to defocus the ghost. Otherwise a different lens configuration, or lens shape change, would need to be tried.

14.4 Thin Lens Ghost Images

Let us consider a thin lens in air and the surfaces contributing a two-bounce ghost image by Fresnel reflections. For an object at infinity, the slope of the marginal ray after passing through the lens including the two surface reflections is,

$$u' = -\phi y \left(3 + \frac{2}{n-1} \right) \approx -7\phi y. \tag{14.2}$$

The marginal ray slop, u', is independent of lens shape, and is seven times larger, for $n = 1.5$, than for the nominal lens, which is $u' = -\phi y$. Thus, the ghost of a thin lens takes place near the lens, and has a low $F/\#$ in comparison to the nominal lens $F/\#$. Figure 14.4 shows ghost images for three shapes of a positive lens.

14.5 Total Internal Reflection Ghost

We have considered two-bounce ghost images with Fresnel surface reflections. It is also possible to have ghost images that take place with a single Fresnel reflection and a total internal reflection (TIR), as shown in Figure 14.5.

Images that result from TIR and Fresnel reflections can be bright because there is only light loss from the single Fresnel reflection. Such ghost images result from bright objects, often outside the field of view of the lens system, and can be avoided by including lens hoods.

14.6 Narcissus Retro-Reflections

As shown in Figure 14.6, light that is reflected from the image plane where there is a light sensor can be reflected back to the image plane to create a ghost image that is in focus with the nominal image. This ghost image effect is known as a narcissus retro-reflection. A metric for narcissus retro-reflections is the first-order product, niy. Whenever there is a surface at an image, $y = 0$, or the marginal ray is concentric with a surface, $i = 0$, a narcissus ghost can take place. This effect increases as the lens system becomes telecentric in image space. For infrared systems, narcissus ghosts are a significant problem.

14.7 Parallel and Concentric Surfaces

A source of ghost images are contiguous surfaces that are concentric, or flat, as in an optical window or beam splitter. Light reflected from the rear surface is

Figure 14.5 Ghost formed with light from an off-axis object entering a lens and being totally internally reflected by the lens second surface and Fresnel reflected by the lens first surface.

Figure 14.6 Telecentric lens suffering from narcissus, due to the front lens surface being concentric with the on-axis light beam. The off-axis beam at 5° is focused in the image plane, then cat's-eye reflected back to the lens where the lens front surface retro-reflects it toward the image plane, creating a ghost image.

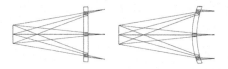

Figure 14.7 Left, ghost images from a parallel plate; right, ghost images from concentric surfaces.

reflected by the front surface to follow the path of the nominal light beam and create a ghost image that is nearly in focus with the nominal image. Some cases are illustrated in Figure 14.7.

Further Reading

Abd El-Maksoud, Rania H., Sasián, José. "Modeling and analyzing ghost images for incoherent optical systems," *Applied Optics*, 50 (2011), 2305–15.

Fest, E. *Stray Light Analysis and Control* (Bellingham, WA: SPIE Press, 2013).

Howard, J. W., Abel, I. R. "Narcissus: reflections on retroreflections in thermal imaging systems," *Applied Optics*, 21(18) (1982), 3393–97.

MacLeod, H. Angus. *Thin-Film Optical Filters* (Boca Raton, FL: CRC Press, 2010).

Murray, Allen E. "Reflected light and ghosts in optical systems," *Journal of the Optical Society of America*, 39 (1949), 30–35.

Smith, G. "Veiling glare due to reflections from component surfaces: the paraxial approximation," *Optica Acta*, V18(11) (1971), 815–28.

Weigel, Thomas, Moll, Bob, Beers, Bart J. "Ghost image debugging on a 240-degree fisheye lens," Proceedings of SPIE 2774, Design and Engineering of Optical Systems (1996); doi: 10.1117/12.246708.

15

Designing with Off-the-Shelf Lenses

Fabrication of a lens system may take several weeks or months. This can be objectionable in a project. Sometimes it is possible to design a lens system out of off-the-shelf single lens elements. For sharp imaging, this might be possible if the lens system is slower than about $F/6$ and the field of view is less than about $\pm 12°$.

There are several companies that offer singlet lenses of different focal lengths, shapes, and diameters. However, lens diameters of more than 75 mm progressively become harder to find. Lenses with diameters of 25 mm or 50 mm are common. Most of the lenses are positive in optical power, but a few are negative. In addition, the glass selection of these lenses is limited; for example to BK7 and SF11 glasses.

For designing a lens system with off-the-shelf singlet lenses, one must be familiar with the lenses available in several lens catalogs. If all the lenses have the same diameter, it might be possible to assemble them in a drop-in lens barrel. It is also possible to integrate, in an off-the-shelf design, lenses that are already well corrected. This chapter provides some examples of lens design with off-the-shelf lenses.

15.1 Cooke Triplet Lens

A choice for designing a lens system with off-the-shelf lens elements can be a Cooke triplet lens. The starting point is then a Cooke triplet lens that meets the lens specifications. Then each of the lenses in the Cooke triplet is divided into two plano lenses, either plano-concave and/or plano-convex. If the radii of curvature of a lens element in the Cooke triplet lens system are similar, then the lens might be adjusted to have both radii the same, and have a double equi-convex or double equi-concave lens, so that there is no need to split the lens.

171

Table 15.1 *Lens prescription for off-the-shelf lens,* f′ = *71.47 mm, FOV =* ±*12°,* F/7 *(mm)*

Surface	Radius	Thickness	Glass	Catalog #
1	18.11	7.01	N-BK7	EO 45146
2	Plano	1		
3	Plano	2.5	C79–80	EO 48322
4	34.38	6.2611		
Stop	Plano	3.5	N-SF11	EO 45020
6	23.54	14.142		
7	40.42	5.3	N-BK7	EO 45296
8	−40.42	46.86		

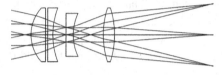

Figure 15.1 Off-the-shelf lens system designed from a Cooke triplet lens. f' = 71.47 mm, FOV = ±12°, *F*/7

The next step is to substitute with an off-the-shelf lens, the lens element that has the strongest optical power. The lens system is re-optimized to restore image quality. The variables used are the curvatures and air spaces of the remaining unfitted lenses. Then the next lens with the strongest optical power is substituted, and the lens system is re-optimized again. This process is repeated until all the plano lenses have been replaced with off-the-shelf lens elements. Airspaces and lens scaling can help to adjust or re-optimize the lens.

Figure 15.1 shows a lens system designed with off-the-shelf lens elements. The first three lenses are plano convex/concave, and the fourth lens is equi-convex. The starting design was a Cooke triplet with glasses N-BK7, N-SF11, and N-BK7. However, to obtain a sharp image and to accommodate for off-the-shelf lenses, the first lens was split into two lens elements. The prescription is given in Table 15.1 using lenses from Edmund Optics. The optimization was done in the visible, *F*, *d*, and *C* wavelengths, and the RMS wavefront is 0.07λ over the field of view.

15.2 UV Lens

Lenses for UV applications are generally expensive. Usually fused silica SiO_2 and calcium fluoride CaF_2 are the materials used because they have high light

Table 15.2 *Lens prescription for UV lens, f' = 105 mm, FOV = ±7°, F/8 (mm)*

Surface	Radius	Thickness	Glass	Catalog #
1	−36.0533	7.0	Silica	Custom
2	33.2023	1.5		
3	51.46	5.4	CaF2	EO 47311
4	−51.46	0.5		
5	51.46	5.4	CaF2	EO 47311
6	−51.46	1		
Stop		14.3058		
7	51.46	5.3	CaF2	EO 47311
8	−51.46	1.7328		
9	−91.3901	7.0	Silica	Custom
10	25.2843	79.2136		

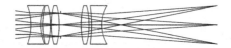

Figure 15.2 UV lens using CaF_2 and SiO_2 materials. $f' = 105$ mm, FOV = ±7°, F/8.

transmission at the UV wavelengths. Calcium fluoride is a delicate material and is difficult to polish well. To reduce cost, the design approach in the design of Figure 15.2 was to use calcium fluoride, off-the-shelf lenses, for the positive elements, and use fused silica, custom made lenses for the negative elements. The lens uses three equi-convex lenses of the same focal length. The constructional data is given in Table 15.2. The field of view (FOV) is ±7°, the speed is F/8, and the focal length is $f' = 105$ mm. The design wavelengths are 300 nm, 400 nm, and 500 nm. The UV lens performance is near diffraction limited.

15.3 Telecentric Lenses

Image space, telecentric lenses with an external stop are useful in optical engineering. They can be combined into optical relays, and they allow for the placement of cube beam splitters and folding mirrors.

Figure 15.3 shows two telecentric lens systems based on an achromatic doublet lens that is off-the-shelf from Edmund Optics, as part EO-45180, $f' = 250$ mm. The doublet specifications are given in Table 15.3. The lens system at the bottom adds a field lens, EO-08074, $f' = -250$ mm, to improve performance over the field of view.

Table 15.3 *Lens prescription for Edmund Optics doublet lens,* f' = *250 mm*

Surface	Radius	Thickness	Glass	Catalog #
1	162.59	9.75	N-BAK4	45180
2	−123.82	3.5	N-SF10	
3	−402.58			

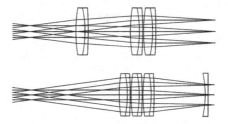

Figure 15.3 (Top) Telecentric lens systems using an off-the-shelf achromatic doublet, EO-45180. (Bottom) The lens system adds a field flattener lens, EO-08074, $f' = -250$ mm.

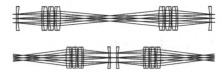

Figure 15.4 Top, image relay lens system. Bottom, pupil relay lens system. Both relays are based on the Edmund Optics achromatic doublet part EO-45180 and a plano concave lens part EO-08074.

The doublet lens is used three times in each telecentric lens. The focal length of the lens system in Figure 15.3 (top) is $f' = 100$ mm, FOV = ±7°, $F/8$. The RMS wavefront error is 0.25λ over the field of view. The focal length of the lens system in Figure 15.3 (bottom) is $f' = 91.2$ mm, FOV = ±12°, $F/8$. The RMS wavefront error is 0.2λ over the field of view. The use of a field flattener lens improves the image quality at the expense of substantially reducing the working distance.

15.4 Relay Systems

Optical relays are often needed in optical engineering. Using twice, front to front and back to back, the telecentric lens system of Figure 15.3 (bottom), two optical relays were designed. Figure 15.4 (top) shows an imaging relay, and Figure 15.4 (bottom) shows a pupil relay system.

15.5 Off-the-Shelf Lens Suppliers

Some suppliers of off-the-shelf lenses are:

1. Edmund Optics, https://www.edmundoptics.com/
2. Ross Optical, http://catalog.rossoptical.com/
3. ThorLabs, https://www.thorlabs.com/

16

Mirror Systems

An important class of optical systems are those that use mirrors. For a mirror, the ray angle of incidence equals the ray angle of reflection, and there is no light dispersion. Using mirrors for imaging has the advantages of allowing for large element diameters, no intrinsic chromatic aberrations, lesser surface curvature for a given optical power, and potential compactness as the beam of light can be folded. The disadvantages are a central obscuration, more sensitivity to surface errors, the need to include baffles to control stray light, and sometimes fewer degrees of freedom to control aberration. Mirror systems, however, use aspheric surfaces to help control aberration. Lenses can be used in conjunction with mirrors to enhance performance. Optical systems that use both mirrors and lenses are known as catadioptric. This chapter discusses some basic mirror systems. The discussion uses aberration coefficients to determine primary aberrations and to find solutions that can later be optimized with real ray tracing.

16.1 Single Mirrors

The design of mirror systems requires finding a mirror layout that meets first-order requirements and that corrects imaging aberrations. To develop the skill of mirror design one first needs to understand simple mirror systems. The use of structural aberration coefficients simplifies aberration assessment and permits trade-off studies; for a single mirror these coefficients are given in Table 16.1.

For an object at infinity, $Y = 1$, and for a parabolic mirror, $K = -1$, of vertex radius, r, with the stop aperture at the mirror, there is no spherical aberration, but there is coma, astigmatism, distortion, and field curvature aberration. A parabolic mirror, as illustrated in Figure 16.1, is used in

Table 16.1 *Structural aberration coefficients of a mirror.*
$K = -\varepsilon^2$ is the conic constant and ε is the eccentricity

Stop at surface	With stop shift
$\sigma_I = Y^2 + K$	$\sigma_I = Y^2 + K$
$\sigma_{II} = -Y$	$\sigma_{II} = -Y(1 - \bar{S}_\sigma Y) + \bar{S}_\sigma \cdot K$
$\sigma_{III} = 1$	$\sigma_{III} = (1 - \bar{S}_\sigma Y)^2 + \bar{S}_\sigma^2 \cdot K$
$\sigma_{IV} = -1$	$\sigma_{IV} = -1$
$\sigma_V = 0$	$\sigma_V = \bar{S}_\sigma \cdot (1 - \bar{S}_\sigma Y) \cdot (2 - \bar{S}_\sigma Y) + \bar{S}_\sigma^3 \cdot K$

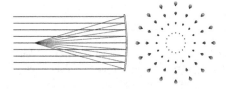

Figure 16.1 Left, a parabolic mirror focusing light from a point object at infinity. The incident beam overlaps with the reflected beam. Right, spot diagrams at the ideal image plane showing coma aberration over a 2° circular field of view.

Newtonian telescopes, where the field of view is mainly limited by coma aberration given by,

$$W_{131} = -\frac{y^3}{r^2}\bar{u}. \tag{16.1}$$

In considering the location of the stop, a conic mirror is free from astigmatism aberration if,

$$\sigma_{III} = (1 - \bar{S}_\sigma Y)^2 + \bar{S}_\sigma^2 \cdot K = 0, \tag{16.2}$$

which requires to have, $1 = \bar{S}_\sigma(Y + \varepsilon)$. Since the structural stop shifting parameter is,

$$\bar{S}_\sigma = \frac{y_P \bar{y}_P \phi}{2K} = \frac{\phi \cdot \bar{s}}{(Y - 1) \cdot \phi \cdot \bar{s} - 2n} = \frac{\phi \cdot \bar{s}'}{(Y + 1) \cdot \phi \cdot \bar{s}' - 2n'}, \tag{16.3}$$

then the entrance pupil location, $\bar{s}$, for a single mirror of radius, r, with the object at infinity, is given by,

$$\bar{s} = \frac{r}{(1 + \varepsilon)}. \tag{16.4}$$

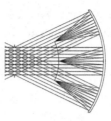

Figure 16.2 A spherical mirror with the stop aperture at its center of curvature. The image surface is spherical.

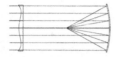

Figure 16.3 Schmidt camera layout. The front refractive element is an aspheric plate that corrects for spherical aberration and that includes curvature to minimize chromatic aberration. The mirror is spherical in shape. To make obvious the aspheric plate curve, an index of refraction of 1.002 was used.

In this case, the entrance pupil coincides with the near focal point of the conic mirror. The structural aberration coefficient for coma aberration is given by,

$$\sigma_{II} = -\varepsilon, \tag{16.5}$$

and the structural coefficient for spherical aberration is,

$$\sigma_I = 1 - \varepsilon^2. \tag{16.6}$$

Thus, a parabolic mirror, $\varepsilon = 1$, with the stop aperture at its front focal point, does not contribute spherical aberration or astigmatism aberration.

For a spherical mirror, $K = 0$, it is possible to avoid coma and astigmatism aberration if the stop is placed at the center of curvature of the mirror, $1 - \bar{S}_\sigma Y = 0$, as shown in Figure 16.2. In this case, the image is limited by spherical aberration and field curvature. A single spherical mirror with the stop at the center of curvature is a highly symmetrical optical system. An optical system where the centers of curvature of all the surfaces are coincident in a point, and the stop aperture is at this point, is called concentric. Such a system has a reduced number of aberrations.

One way to correct for spherical aberration is to include an aspheric plate at the center of curvature of the spherical mirror, as shown in Figure. 16.3. This system is known as the Schmidt camera, which is capable of imaging relatively large fields of view at a fast optical speed. However, the Schmidt camera suffers from field curvature aberration and from obscuration after an image sensor is placed at the focal surface.

Since fourth-order spherical aberration of a spherical mirror is,

$$W_{040} = -\frac{1}{4}\frac{y^4}{r^3}, \tag{16.7}$$

and the spherical aberration of an aspheric plate in air is, $W_{040} = (n-1)A_4y^4$, then the fourth-order aspheric coefficient of the Schmidt correcting plate is given by,

$$A_4 = -\frac{1}{4(n-1)}\frac{1}{r^3}. \tag{16.8}$$

A radius of curvature is also specified as part of the aspheric surface to minimize chromatic aberration from the plate, and to reduce the slope of the surface near its edge.

16.2 Two-Mirror Systems

Two-mirror combinations allow us to have easier access to the image and provide more degrees of freedom to satisfy first-order requirements and to correct aberrations. Two classical configurations are the Cassegrain and the Gregorian telescopes shown in Figure 16.4.

Table 16.2 provides the structural aberration coefficients of a two-mirror system with the stop at the primary mirror. The parameter, L, is the ratio of the mirror separation to the back focal length (distance from the secondary vertex to the image plane). Structural aberration coefficients allow us to perform parametric studies, and provide clues into what to change to control aberration. However, once a solution is known to exist, it is convenient to find it using the optimizer in a lens design program by targeting fourth-order aberration coefficients to zero.

As shown in Table 16.3, depending on the mirror conic constant, there are several solutions for a Cassegrain type configuration free from spherical aberration. In the Cassegrain configuration, both mirrors are independently corrected for spherical aberration; the primary mirror is parabolic, and the

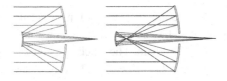

Figure 16.4 Left, The Cassegrain configuration has a convex secondary mirror. Right, The Gregorian configuration has a concave secondary mirror and provides an erect image.

Table 16.2 *Structural coefficients of a two-mirror system. Stop at primary mirror. Object at infinity; m is the transverse magnification of the secondary mirror, and L is the ratio of the mirror separation to the back focal distance*

$$\sigma_I = m^3(1 + K_1) + \frac{(1-m)^3}{1+mL}\left(\left(\frac{1+m}{1-m}\right)^2 + K_2\right)$$

$$\sigma_{II} = -m^2 + (1-m)^2\left(-\left(\frac{1+m}{1-m}\right)\left(1 - \frac{1}{2}\frac{(1-m)L}{1+mL}\left(\frac{1+m}{1-m}\right)\right) + \frac{1}{2}\frac{(1-m)L}{1+mL}K_2\right)$$

$$\sigma_{III} = -1 + (1-m)(1+mL)\left(\left(1 - \frac{1}{2}\frac{(1-m)L}{1+mL}\left(\frac{1+m}{1-m}\right)\right)^2 + \left(\frac{1}{2}\frac{(1-m)L}{1+mL}\right)^2 K_2\right)$$

$$\sigma_{IV} = -m - (1-m)(1+mL)$$

$$\sigma_V = \frac{1}{(1+mL)^2}\left(\left(\frac{1}{2}\frac{(1-m)L}{1+mL}\right)\left(1 - \frac{1}{2}\frac{(1-m)L}{1+mL}\frac{1+m}{1-m}\right)\left(2 - \frac{1}{2}\frac{(1-m)L}{1+mL}\frac{1+m}{1-m}\right) + \left(\frac{1}{2}\frac{(1-m)L}{1+mL}\right)^3 K_2\right)$$

Table 16.3 *Conic constants of two-mirror configurations corrected for spherical aberration*

Configuration	Primary mirror	Secondary mirror
Cassegrain	$K_1 = -1$	$K_2 = -\left(\frac{1+m}{1-m}\right)^2$
Dall-Kirkham	$K_1 = -1 - \frac{(1-m)(1+m)^2}{m^3(1+mL)}$	$K_2 = 0$
Pressman-Carmichel	$K_1 = 0$	$K_2 = -\left(\frac{1+m}{1-m}\right)^2 - \frac{m^3(1+mL)}{(1-m)^3}$
Ritchey-Chretien (aplanatic)	$K_1 = -1 - \frac{2}{Lm^3}$	$K_2 = -\left(\frac{1+m}{1-m}\right)^2 - \frac{2(1+mL)}{L(1-m)^3}$

secondary mirror is hyperbolic. For ease of manufacturing, the secondary mirror can be made spherical, which requires an elliptical primary mirror; this solution is known as Dall-Kirkham. The Ritchey-Chretien system uses hyperbolic mirrors, and corrects for spherical aberration and coma aberration. Indeed, there is a family of solutions depending on the mirrors conic constants, and one member is aplanatic. The Pressman-Carmichel system uses a spherical primary mirror and has a large amount of coma aberration.

For a Cassegrain or Gregorian configuration, the structural coefficient for coma becomes, $\sigma_{II} = -1$. Since this coincides with the structural coefficient of a single mirror, such as a paraboloid, then the coma of a Cassegrain or Gregorian system is the same as the coma of a paraboloid mirror of the same focal length.

16.3 Spherical Mirror Solutions

If both mirrors are spherical and the magnification of the secondary is $m = \frac{-1 \pm \sqrt{5}}{2}$, then spherical aberration, coma, and astigmatism are corrected. The first solution, $m = \frac{-1 + \sqrt{5}}{2}$, has the primary mirror concave, the secondary mirror convex, and a virtual image. The second solution, $m = \frac{-1 - \sqrt{5}}{2}$, has a convex primary mirror, a concave secondary mirror, and a real image, as shown in Figure 16.5. In this solution, both mirrors are concentric. Some microscope objectives for use in the UV region of the spectrum have been designed and built using this two-mirror concentric solution.

16.4 Schwarzschild Flat-Field, Anastigmatic Solution

When both mirrors have the same but opposite vertex radius of curvature, there is a solution free form spherical aberration, coma, astigmatism, and field curvature, as shown in Figure 16.6. The primary mirror is convex, and the

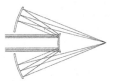

Figure 16.5 Concentric spheres, free from spherical aberration, coma, and astigmatism.

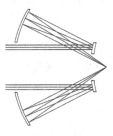

Figure 16.6 Schwarzschild flat-field anastigmatic two-mirror solution.

secondary mirror is concave. However, this configuration is not very practical because of the large mirror perforations, beam obscuration, and stray light reaching the image plane directly.

16.5 Mersenne Telescopes

For two mirrors that are arranged to be afocal, as shown if Figure 16.7, the Seidel sums are given in Table 16.4. If both mirrors are paraboloids, then the combination is free from spherical aberration, coma, and astigmatism. These solutions are known as the Mersenne telescopes.

16.6 Paul and Paul–Baker Systems

Design flexibility for correcting aberration to achieve large fields of view at a fast optical speed can be gained by adding more mirrors to a system.

Table 16.4 *Seidel sums for two-mirror afocal systems. Stop aperture at primary mirror.*
$\mathcal{K} = 1, \phi_1 = 1, y_1 = 1, \bar{y}_1 = 0, y_2 = m$

$$S_I = \frac{1}{4}((1 + K_1) - m(1 + K_2))$$

$$S_{II} = \frac{1}{4}(m - 1)(1 + K_2)$$

$$S_{III} = -\frac{1}{4}\frac{(m - 1)^2}{m}(1 + K_2)$$

$$S_{IV} = -\frac{m - 1}{m}$$

$$S_V = \frac{1}{4}\frac{m - 1}{m^2}\left(8 + 6(m - 1) + (m - 1)^2(1 + K_2)\right)$$

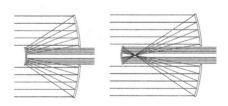

Figure 16.7 Two Mersenne telescopes using parabolic mirrors. They are free from spherical aberration, coma, and astigmatism.

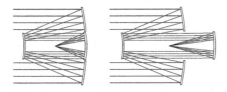

Figure 16.8 Left, Paul three-mirror system. Right, Baker-Paul three-mirror system.

The Mersenne telescope can be made focal by adding a third spherical mirror, as shown in Figure 16.8 (left). The stop aperture is located at the secondary mirror, as well as the center of curvature of the tertiary spherical mirror. The tertiary mirror has the same radius of curvature as the secondary mirror, but opposite in sign, and contributes spherical aberration and field curvature. The spherical aberration contributed by the tertiary mirror is,

$$W_{040} = -\frac{1}{4}\frac{y^4}{r^3},$$
(16.9)

and can be corrected with an aspheric reflecting plate of asphericity,

$$A_4 = \frac{1}{8}\frac{y^4}{r^3}.$$
(16.10)

If this plate is combined with the parabolic secondary mirror of the Mersenne telescope, which has the opposite asphericity, the secondary mirror becomes spherical. Effectively, the Mersenne telescope acts as a corrector for the spherical tertiary mirror. The result is a system that has a parabolic primary mirror, spherical secondary and tertiary mirrors, and that is free from spherical aberration, coma, and astigmatism.

The radius of curvature of the tertiary spherical mirror can be chosen to correct for Petzval field curvature of the Mersenne telescope, and, by making the secondary mirror elliptical in shape, the system becomes flat-field and anastigmatic, as shown in Figure 16.8 (right).

16.7 Offner Unit Magnification Relay

Optical imaging relays are an important class of systems. Three spherical concentric mirrors, as shown in Figure 16.9, provide telecentricity in object and image spaces at unit negative magnification, and are free from the five monochromatic aberrations. In practice, this imaging relay is used as a ring, or annular, field system in that the usable field portion is off-axis and annular.

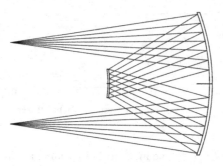

Figure 16.9 Offner unit magnification relay with two concentric spherical mirrors and three light reflections.

Figure 16.10 Meinel two-stage configuration for a large deployable telescope.

The primary and tertiary mirrors can be portions of a single large mirror, and the system is then a two mirror, three reflection one. The stop aperture is placed at the convex secondary mirror, and odd aberrations are corrected by symmetry. Since the radius of curvature of the primary and tertiary mirrors is twice that of the secondary, but opposite in sign, Petzval field curvature is corrected. None of the mirrors contributes spherical aberration, because the refraction invariant, $A = ni = n'i' = 0$, is zero for the three mirrors. Finally, in a doubly telecentric system, the pupil astigmatism equals the image astigmatism, $\bar{W}_{222} = W_{222}$, and, since the pupil astigmatism is zero because $A = 0$ for the three mirrors, then the image astigmatism is also zero.

16.8 Meinel Two-Stage Telescope

In order to correct for aberrations caused by errors in a large space-deployable telescope, a four-mirror configuration, known as a Meinel two-stage telescope, can be designed, as shown in Figure 16.10. The first two-mirror stage is formed by a Cassegrain type telescope. The second two-mirror stage relays the image formed by the first stage to the final location. The second stage has

Table 16.5 *Prescription for a four-mirror telescope with spherical primary and secondary mirrors. The stop aperture is at the primary mirror and 5.0 meters in diameter.* f' = 59,310 mm; F/12

Mirror	Radius, mm	Thickness, mm	Index	A_4
1	−18,000.00	−7,500.00	−1	
2	−4,500.00	9,080.00	1	
3	−6,932.98	−4,960.00	−1	7.9501×10^{-14}
4	44,575.77	6,560.00	1	-1.15×10^{-10}

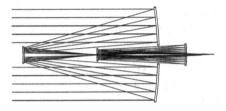

Figure 16.11 Flat-field, aplanatic/anastigmatic four-mirror configuration for large telescopes.

the quaternary mirror coinciding with the exit pupil of the system. The tertiary mirror makes an image of the exit pupil of the first stage on the quaternary mirror. Optically, the quaternary mirror is equivalent to the large primary mirror, in that field independent aberrations from the primary mirror can be corrected at the quaternary mirror.

If the primary mirror is spherical, then the quaternary mirror is strongly aspheric to correct for spherical aberration. Because coma aberration, W_{131}, contributed by an aspheric surface depends on the ratio, $\bar{y}/y$, at the quaternary mirror, and on the amount of asphericity,

$$W_{131} = 4\left(\frac{\bar{y}}{y}\right)A_4 y^4 \Delta(n), \qquad (16.11)$$

then a slight axial repositioning of the quaternary mirror can be used to generate positive or negative coma and make the system aplanatic, as shown in Figure 16.11. Thus, a two-stage mirror configuration can be aplanatic with only one strongly aspheric mirror. If more mirrors are made aspheric, all five monochromatic aberrations can be corrected. A prescription for a flat-field and anastigmatic solution to fourth-order is given in Table 16.5. Note that, after reflection, the sign of the distance to the next mirror and the sign of the index of refraction changes.

Further Reading

Baker, J. G. "On improving the effectiveness of large telescopes," *IEEE Transactions on Aerospace and Electronic Systems*, AES-5(2) (1969), 261–71.

Brueggemann, H. P. *Conic Mirrors* (London: The Focal Press, 1968).

Korsch, Dietrich. *Reflective Optics* (San Diego: Academic Press Inc., 1991).

Offner, A. "New concepts in projection mask aligners," *Optical Engineering*, 14(2) (1975), 142130.

Paul, Maurice. "Systèmes correcteurs pour réflecteurs astronomiques," *Revue d'Optique Theorique et Instrumentale*, 14(5) (1935), 169–202.

Sasián, J. "Flat-field, anastigmatic, four-mirror optical system for large telescopes," *Optical Engineering*, 26(12) (1987), 1197–99.

Schwarzschild, K. "Untersuchungen zur geometrischen Optik II," *Astronomische Mittheilungen der Koeniglichen Sternwarte zu Gottingen*, 10 (1905), 3–27.

Shafer, David R. "Anastigmatic two-mirror telescopes: some new types," *Applied Optics*, 16 (1977), 1178–80.

Stacy, J. E., Meinel, A. B., Meinel, M. P. "Upgrading telescopes by active wavefront pupil correction," in 1995 International Lens Design Conference, Proceedings of SPIE, 554 (1986), 186–90.

Wilson, R. N. *Reflective Telescope Optics I* (Berlin: Springer-Verlag, 1996).

17

Miniature Lenses

There are a number of technological applications that use miniature lenses in which the lens diameter is a few millimeters, and typically smaller than 10 mm. For lens systems that employ such miniature lenses, several advantages result because of the scale. For a given lens form and, except for distortion, the aberrations scale down, while the wavelength of light remains the same. Given that lens volume is small, a wider possibility of lens materials becomes possible due to cost or material limitations. Lens weight is reduced, as well as dimensional changes due to temperature. Further, very thick lenses can be used in some applications. However, lens tolerances for lens thickness and decenter become tighter. Many microscope objectives, lenses for endoscopes, and lenses for mobile phones are in the category of miniature lenses. This chapter provides a discussion about lens design for mobile phones lenses.

17.1 Lens Specifications

Lenses for mobile phones, as illustrated in Figure 17.1, have been developed over the last two decades. Lens designs have evolved from having one or two lens elements, to three and four elements, to five-to-eight lens elements. Some of the lens elements have been made by glass molding, but currently they are made by plastic injection molding. To aid in the correction of aberration, highly aspheric surfaces are used. Table 17.1 provides typical lens specifications for field of view (FOV), focal length, F-number, total track length (TTL) from the vertex of the first surface of the lens to the image plane, chief ray angle of incidence at image sensor (CRA), relative illumination (RI), and number of lens elements.

Like any other photographic lens, the design of a mobile phone lens is driven by the light sensor; for example, a Charge Coupled Device (CCD) or

Table 17.1 *Typical mobile phone lens specifications*

Year	2006	2012	2018
Focal length	3–6 mm	3–5 mm	3–5 mm
FOV	66°	72°	78°
F/#	2.8	2.2–2.4	2.0–1.4
TTL	<5.0 mm	<5.0 mm	<6.0 mm
Distortion	<1–2%	<1–2%	<1–2%
CRA	<24°	<30°	<33°
RI	>50%	>50%	>32%
# lens elements	3–5	4–6	5–8

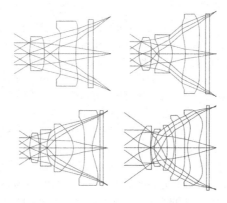

Figure 17.1 Mobile phone lens forms with two, three, four, and five lens elements. The plane parallel plate next to the image plane represents an infrared filter.

Complementary Metal Oxide Semiconductor (CMOS) image sensor. The diagonal of the active area of the sensor defines the minimum image circle diameter; some allowance in this diameter is given to allow for some lens/ sensor decenter. There are many format sizes for an image sensor that are given in inches, such as 1/3″. This format has a diagonal of 6 mm, a width of 4.8 mm, and a height of 3.6 mm. The inch designation of image sensors is an arcane reference to the vidicon tubes used in early video cameras. It is best to look up the actual digital image sensor dimensions for a given sensor specification.

The lens design is also driven by the specified total track length, which is often required to be less than 6 mm, so that the lens can be integrated in a thin mobile phone. Although most designs for mobile phone lenses are not telephotos, the ratio of the total track lens to the focal length, the telephoto ratio, is still used. A typical value for this ratio is 1.2.

Table 17.2 *Example of MTF specifications in fractions of the Nyquist frequency and in cycles/mm for an object at infinity*

	N_Q	$N_Q/2$	$N_Q/4$
MTF On-axis	>40%	>60%	>80%
MTF Off-axis @ 0.7 field	S>30%	S>50%	S>70%
	T>20%	T>40%	T>60%
MTF Off-axis @ 1.0 field	S>20%	S>30%	S>40%
	T>10%	T>20%	T>30%

Given the focal length and the image circle diameter, and assuming no distortion, the field of view can be calculated. Thus, for a focal length of 4.8 mm and an image circle diameter of 6 mm, the field of view is $\pm 32°$.

Electronic sensors have pixels sensitive to the light incident on them. Typical pixel sizes for CMOS sensors vary in the range of 0.001 mm to 0.008 mm. A given spatial frequency can be recovered whenever it is sampled with twice that frequency. This sampling frequency is known as the Nyquist frequency, N_Q. An image sensor can sample an image at a spatial frequency of 1/(pixel size). It follows that the sensor can recover a spatial frequency, N_Q = 1/(2 × pixel size). If the pixel size is 1.5 μm, then N_Q = 333 cycles/mm (also line pairs per mm; lp/mm). The maximum frequency that a lens with a circular aperture can image is N_C = 1/($\lambda F/\#$). Using λ = 0.5 μm and $F/2.2$ yields N_C = 909 cycles/mm. Thus, a diffraction limited lens at $F/2.2$ would not limit the sensor spatial frequency sampling. The image quality for a lens can then be specified in terms of the Nyquist frequency, for different fields and object distances, as shown in Table 17.2.

The design of a mobile phone lens also requires consideration of the spectral bandwidth to be used, for example, the visible spectrum from 400 nm to 700 nm. A filter is used in mobile phone lenses to suppress infrared radiation (IR). This filter is modeled as a parallel plate of BK7 glass as the last element of the lens system. The filter introduces spherical aberration that must be compensated by a lens element near the stop aperture. In addition, the spectral response of the sensor is used to weight the wavelengths used by the lens optimizer to reflect the sensor's sensitivity to wavelength.

A CMOS sensor includes an array of micro-lenses, also called lenslets, placed on top of the light sensitive pixels. As shown in Figure 17.2, the function of each micro-lens is that of a field lens that forms an image of the exit pupil of the mobile phone lens onto the surface of each light sensitive element of a pixel. In this way, light is redirected to the active areas in each pixel, as not all the area of a pixel is light sensitive. The array of micro-lenses

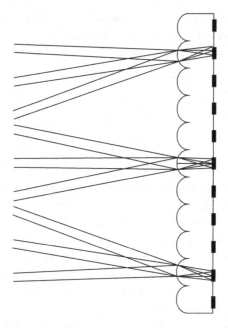

Figure 17.2 Model of array of micro-lenses to improve light coupling efficiency. The micro-lenses act as field lenses, forming a pupil at the CMOS light sensitive elements.

requires a specific chief ray angle (CRA) of incidence as a function of the ray height on the sensor. The specification can be simply to match a given CRA at the corner of the sensor, or to match within one or two degrees a table of angles of incidence as a function of the radial position on the sensor.

There must be enough back focal length (BFL) to allow placement of the image sensor and any clearances required by opto-mechanical considerations.

17.2 Lens Design Considerations

Mobile phone lenses have been evolutionary, in that every generation increased complexity from the previous one. The first-order layout in practice is usually taken from an existing lens design form. Given that lens element axial thickness can be substantial, thick lenses can be used, and more design forms are possible whenever the lens system has two or three lens elements, or when the total track length is not a concern. Many combinations of optical power for the lens elements are possible, for example PNPN, PNNP, NPPN for a four-lens element system. The patent literature has hundreds of lens

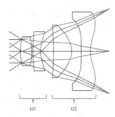

Figure 17.3 Four-lens element mobile phone lens. The first and second lens elements form the front group, and the third and fourth lens elements form the rear group.

design examples for mobile phone lenses and their forerunners, personal digital assistants.

Figure 17.3 shows a four-element mobile phone lens divided into two groups, G_1 and G_2. The stop aperture is placed in front to reduce the chief ray angle of incidence at the image sensor, or slightly inside the lens to help provide some lens symmetry and to reduce odd aberrations. However, provided the relative illumination specification is met, the stop aperture can be placed in front, and some vignetting by an aperture inside the lens can be allowed. This vignetting has the effect of making the stop aperture appear inside the lens for off-axis field points. For stray light control, several apertures are usually specified within the mobile phone lens assembly.

Significant optical power of the lens system is in the front lens group, which must provide the majority of the correction for chromatic aberration, spherical aberration, and coma. The rear group corrects for Petzval field curvature, astigmatism, and distortion aberration. This group usually has highly aspheric surfaces. In particular, the last aspheric surface changes the optical power, from negative near the optical axis to positive near the edge of the lens. This positive power helps to decrease the CRA and introduces positive pupil coma, which helps to increase the relative illumination. Pupil spherical aberration is also present due to the positive power of the last lens near the edge.

The total track of the lens system has a strong impact on imaging quality. The smaller the telephoto ratio is, the more difficult it becomes to meet image quality requirements, in part because the lens is optically stressed and higher order aberrations become larger.

It is good practice not to specify unnecessary aspheric terms and to keep them at a minimum number. If many aspheric terms are used, the lens optimizer may create surface features in one lens that are canceled in another lens. Under lens decenter, the features can partially add rather than cancel, with the result that the as-built-lens becomes more sensitive to manufacturing errors. Lens elements near the stop aperture may need only two or three aspheric terms, as their function is mainly to correct for spherical aberration.

As mentioned before, one problem with the small scale of miniature lenses is that lens element thickness and decenter errors can have a large impact by decreasing performance. For example, a thickness error of 0.1 mm can be tolerable in a 50 mm focal length lens, but not at all in a miniature lens with a focal length of 5 mm. Therefore, an important part of the lens design is to desensitize as much as possible a given design. The wavefront and distortion correction should not have rapid changes at the edge of the aperture or field, or slightly beyond the image circle, that could substantially decrease the performance of the as-built lens. Minimizing ray angle of incidence over all the lens surfaces helps to make the lens system less sensitive to errors.

The first and second lens elements in the front group usually have strong optical power to correct for chromatic aberration and provide most of the system's optical power. Any manufacturing or assembling errors in these lens elements can introduce significant aberration that would degrade the image sharpness. Field correcting elements in the rear group have a small beam footprint and a large chief ray height, and, under manufacturing errors, they can introduce asymmetrical image distortion.

During the early stages of the design of a mobile phone lens, and to explore possible lens forms, the surface conic constant and the fourth order aspheric coefficients can be released as variables for optimization. Once a promising design form is found, more aspheric terms can be added, provided they contribute to improve performance. It is important that, for optimization and evaluation, sufficient pupil and field sampling points be specified. Typically, the field is sampled with at least ten field points.

The optimization of a mobile phone lens can start with minimizing RMS spot size, then minimizing RMS wavefront error, and finally adjusting the lens to meet MTF requirements. If a given design form cannot meet image quality specifications, then a lens designer may consider changing lens materials or adding one more lens element. To add a lens element, a parallel plate of zero thickness is inserted in an air space, or a thick lens is split into two lenses with no air space between them. Fourth-order aspheric terms are included in the new surfaces, and the system is re-optimized, not necessarily allowing the radii of the new surfaces to vary. Then thickness is added to the new surfaces by small increments until a lens that is physically possible, this is with positive thicknesses and non-overlapping lenses, is obtained. As lens complexity is added, more design forms are possible, and lens image quality is expected to improve. The lens form of Figure 17.3 has been used as a starting point in many patented lenses to develop designs with more lenses. Parallel plates of

Table 17.3 *Example of a four-lens element design for a mobile phone lens.*
f' = 5.0 mm, FOV = ±32°, TTL/f = 1.4, F/2.8, CRA = 30.6°

Surface	Radius	Thickness	Plastic	K	A_4
STOP					
2	3.5432.72	0.87	E48R	−3.2457	
3	−4.57977	0.1		−11.7213	
4	126.2449	0.6	OKP4	0.0	
5	3.013807	0.8		−0.9977	-2.533×10^{-3}
6	−8.81693	1.61	E48R	0.0	4.161×10^{-3}
7	−1.61409	1.0		−2.2699	−0.0126713
8	72.73421	0.7	OKP4	0.0	−0.0100888
9	2.101288	0.7		−6.3020	-6.073×10^{-3}
10	Plano	0.3	BK7		
11	Plano	0.4			
Image					

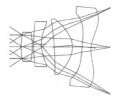

Figure 17.4 Example of a lens system with exaggerated lens element decenter and tilt.

glass have been added and then optimized to obtain five, six, and seven element designs.

Lens desensitizing to manufacturing errors is still an important requirement, so that the as-built lens meets fabrication yield specifications. Figure 17.4 shows exaggerated lens decenter and tilt, which can happen during lens assembly. Under lens element decenter and tilt, the axial symmetry is broken, and new aberration forms can appear. For example, in addition to uniform spherical aberration, uniform coma and uniform astigmatism can take place. In addition to linear coma, linear astigmatism and field tilt can also take place. Further, in addition to cubic distortion, asymmetric quadratic distortion can take place. Here uniform refers to the aberration being independent of the field of view, linear as dependent linearly with the field, and so forth. Appendix 4 provides a table of aberrations that can happen when the lens symmetry is reduced from axial to plane symmetry.

An example of a four-lens element design is given in Table 17.3 and shown in Figure 17.1 (bottom left). In this design, only conic constants and fourth-order coefficients of asphericity were used. This is to emphasize that the use of many higher order aspheric coefficients, such as up to sixteenth order, may not be necessary as they can be redundant.

It is common to have errors in the aspheric coefficients due to truncating numbers, incorrect algebraic signs, neglecting numbers, or swapping coefficients. Thus, typically a lens, say from the patent literature, would need to be re-optimized to correct errors, to reflect actual indices of refraction, to adjust focal length, field of view, optical speed, MTF performance, and to reflect current manufacturing requirements such as minimum central and edge thicknesses.

17.3 Lens Manufacturing Considerations

Early lens systems used glass for the first lens element to provide a large v-number difference for the correction of chromatic aberration, and for making the lens system less sensitive to environmental changes such as temperature and humidity variations. However, with the development of plastic materials with low v-numbers, low birefringence, low water absorption, and the intro-duction of auto-focus, lens systems with all the lens elements made by plastic injection molding have been realized.

Under large volume production, lenses made by plastic injection are pre-ferred to lenses made by glass molding, because of cost. Some advantages of plastic lens molding are the freedom to specify aspheric surfaces, and the choice to specify a lens flange to help precisely position a lens element with respect to other lens elements, thereby simplifying the lens system assembly.

For proper plastic flow and cooling, plastic lens manufacturers have some requirements for the aspect ratio of positive and negative lenses. Some guide-lines are as follows: for positive lenses the ratio of lens central thickness to edge thickness should not be more than 3.2, and the edge thickness should not be less than 0.32 mm; for negative lenses the ratio of the maximum thickness to the central thickness should not be larger than 2.7, and the central thickness should not be less than 0.27 mm. These requirements are over the clear aperture of the lens and do not consider the lens flange; they are intended for the lens designer. Further, they also depend on the capabilities of the manufac-turer. The ideal lens for injection molding approaches a lens with parallel surfaces so that plastic flow, cooling, and shrinkage are uniform. For miniature lenses, tolerances of 10 µm for thickness, decenter, total indicator runoff, and lens tilt are low; tolerances between 2 µm and 5 µm are medium and feasible to achieve in manufacturing, and tolerances between 0.5 µm and 2 µm are challenging to achieve, and are met for some miniature optics.

Table 17.4 *Properties of some plastics used in mobile phone lenses*

Code	n_d	v	γ	ρ
480R	1.525	55.95	$+1.44 \times 10^{-4}$	1.01
E48R	1.531	56.04	-2.62×10^{-4}	1.02
F52R	1.534	57.09	-2.21×10^{-4}	1.01
OKP4	1.607	26.90	-3.44×10^{-4}	1.20
OKP4HT	1.632	23.33	-2.72×10^{-4}	1.24

n_d, index of refraction; v, v-number; γ, opto-thermal coefficient; ρ, specific gravity.

The optical and mechanical properties of some plastics used in miniature lens systems are shown in Table 17.4. Some advantages of plastics over glass are that they are moldable at a lower temperature, have lower cost for large volumes, with lower weight, and aspheric and diffractive surfaces can be specified. Some disadvantages are the greater sensitivity to temperature changes, increased sensitivity to water absorption, internal light scattering or haze, reduced light transmission below 450 nm and above 1,000 nm, and a low resistance to abrasion.

Further Reading

Bareau, Jane, Clark, Peter P. "The optics of miniature digital camera modules," Proceedings of SPIE 6342, International Optical Design Conference 2006, 63421F (2006), doi: 10.1117/12.692291.

Clark, P. "Lens design and advanced function for mobile cameras," Chapter 1, in *Smart Mini-Cameras*, Galstian, T. V., ed. (Boca Raton, FL: CRC Press, 2014).

Clark, Peter P. "Mobile platform optical design," Proceedings of SPIE 9293, International Optical Design Conference 2014, 92931M (2014), doi: 10.1117/12.2076395.

Reshidko, D., Sasián, J. "Optical analysis of miniature lenses with curved imaging surfaces," *Applied Optics*, 54(28) (2015), E216–23.

Schaub, M. *The Design of Plastic Optical Systems* (Bellingham, WA: SPIE Press, Vol. TT80, 2009).

Sure, Thomas, Danner, Lambert, Euteneuer, Peter, Hoppen, Gerhard, Pausch, Armin, Vollrath, Wolfgang. "Ultra-high-performance microscope objectives: the state of the art in design, manufacturing, and testing," Proceedings of SPIE 6342, International Optical Design Conference 2006, 63420E (2006), doi: 10.1117/12.692202.

Yan, Yufeng, Sasián, José. "Miniature camera lens design with a freeform surface," Proceedings of SPIE 10590, International Optical Design Conference 2017, 1059012 (2017), doi: 10.1117/12.2292653.

Links to optical plastics vendors:
http://www.ogc.co.jp/e/products/fluorene/okp.html
https://www.zeonex.com/Optics.aspx.html#glass-like

18

Zoom Lenses

By moving groups of lenses along the optical axis of a lens system, it is possible to continuously vary the focal length of the system, which results in a varifocal system. A zoom lens results whenever the position of the image plane remains stationary. As the focal length of the lens changes, the field of view also changes. The focal length is varied by moving axially at least one group of lenses, called the variator. To maintain the image plane position as stationary, another group of lenses is required, which is called the compensator. The axial movement of the variator and the compensator are usually different in nature. The variator might be moved in a linear manner, and the compensator in a non-linear manner by using a mechanical cam. As shown in Figure 18.1, zoom lenses that maintain the image position by moving the variator and the compensator equally are referred to as having optical compensation. Zoom lenses that require different movements for the variator and the compensator are referred to as mechanically compensated. The variator and the compensator constitute the lens kernel of the zoom lens. Mechanically compensated lenses have more optical design freedom than optically compensated lenses, and most modern zoom lenses are of the former class.

The ratio γ of the maximum focal length to the minimum focal length achieved by a zoom lens is known as the zoom ratio or range. A small zoom ratio is in the range of 1–3, a medium zoom ratio is between 3 and 12, and a large zoom ratio is larger than 12. The zoom ratio is an important zoom lens parameter because the larger it is the more complex the zoom lens becomes for a given performance. In the context of zoom lenses, a lens group, sometimes also referred to as a lens unit, consists of consecutive lenses with their respective distances fixed. Thus, for example, there are fixed groups, moving groups, focusing groups, variator groups, and compensator groups. In other lens design contexts a lens group is defined according to the main optical function it performs. For example, a focusing group and a field correcting group.

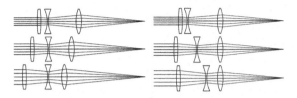

Figure 18.1 Left, Optically compensated zoom lens. The compensator and var-iator (first and third lenses) are mechanically coupled and move in a single movement. Right, Mechanically compensated zoom lens. The compensator and variator (first and second lenses) move differently.

Some earlier zoom lenses consisted of an afocal lens capable of changing its transverse magnification, followed by a standard imaging lens known as the prime lens. The aperture stop was located at the prime lens, or after the last moving group, and, as the magnification of the afocal lens was changed, the $F/\#$ of the combination remained the same. This was a desirable feature for film exposure. With the advent of automatic exposure, where the stop diameter was varied to maintain the exposure while zooming, the position of the stop was no longer required to be after the last moving group, and it could be part of the variator or compensator lens groups. However, the stop position in a zoom lens plays a role in controlling the overall lens diameter and light vignetting. The aberration function $W(\vec{H}, \vec{\rho}, \gamma)$ of a zoom lens can be written as

$$W\left(\vec{H}, \vec{\rho}, \gamma\right) = W_P\left(\vec{H}, \vec{\rho}\right) + W_K\left(\vec{H}, \vec{\rho}\right) + W_K\left(\vec{H}, \vec{\rho}, \gamma\right), \qquad (18.1)$$

where $W_P(\vec{H}, \vec{\rho})$ is the aberration function of the prime lens, or of non-moving lens groups, $W_K(\vec{H}, \vec{\rho})$ is the aberration function of the kernel that is independent of the zoom ratio γ, and $W_K(\vec{H}, \vec{\rho}, \gamma)$ is the aberration function of the kernel that depends on the zoom ratio. In part, the design of the kernel involves finding a variator and a compensator that renders $W_K(\vec{H}, \vec{\rho}, \gamma)$ tolerable, or negligible, over the zooming range. In simple lenses, $W_K(\vec{H}, \vec{\rho}, \gamma)$ is made negligible at an intermediate zoom position and tolerable at the zoom range extremes. In other zoom lenses, $W_K(\vec{H}, \vec{\rho}, \gamma)$ is made negligible at two or more positions of the zooming range, and tolerable at the remaining positions. The kernel aberrations, $W_K(\vec{H}, \vec{\rho})$, that are independent of the zoom ratio, are balanced or corrected by the aberrations, $W_P(\vec{H}, \vec{\rho})$, of the non-moving lens groups. By breaking the aberration function of a zoom lens into aberrations that depend, or do not depend, on the zoom ratio γ, the design task is conceptually and practically simplified. The lens design of zoom systems is quite interesting, and notably zoom kernels can be quite simple.

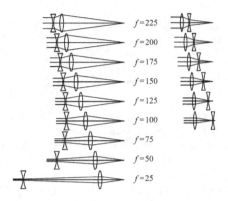

Figure 18.2 Left, Two-group reverse telephoto zoom. Right, Two-group tele-photo zoom. The focal length of the groups is ±100.

18.1 Two-Group Zoom

The optical power, ϕ, of the combination of two optical systems, A and B, in air is given by,

$$\phi = \phi_A + \phi_B - t\phi_A\phi_B. \tag{18.2}$$

By changing the spacing, t, between the systems, the optical power of the combination can be changed. There are four possibilities for the optical power of the individual lens systems, or groups, and these are positive-positive (PP), negative-negative (NN), positive-negative (PN), and negative-positive (NP). Figure 18.2 shows how the groups move as a function of the focal length for the reverse telephoto (NP) and for the telephoto (PN) cases. Across the zooming range, the variator moves nearly linearly, while the compensator moves non-linearly.

The telephoto PN zoom lens has the advantage of compactness, but the disadvantages of limited field of view and zoom ratio. The reverse telephoto PN zoom lens has the advantages of providing a large field of view, favorable aberration compensation, a large zoom ratio, and a large back focal length. However, the reverse telephoto length is larger than the length of the telephoto lens.

Two-group zoom lenses have restricted design freedom because, for an object at infinity, they must operate with collimated incident light, and they must also provide focused light. Thus, the choice of optical power of the variator and compensator is restricted. In contrast, zoom lenses with lens groups before and after the zoom kernel allow the variator and compensator groups to receive and deliver either convergent or divergent light. It is,

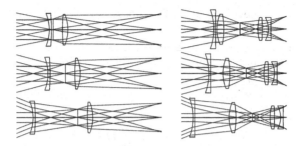

Figure 18.3 Left, Two-group zoom lens partially corrected for fourth-order aberrations at the zooming range extremes. Right, two-group, four-lens zoom lens corrected for fourth-order aberrations at the zoom range extremes.

however, quite notable that two-group zoom lenses can meet first-order requirements and provide sharp images.

The design of a zoom lens may start by a first-order layout to determine group optical power and to allow for enough clearance so that the variator and the compensator do not overlap with themselves or overlap with other lens groups. The first-order layout should satisfy first-order requirements, including the zooming range.

18.2 Example

Figure 18.3 (left) shows a two-group zoom lens corrected for fourth-order spherical aberration, astigmatism, and field curvature at the extremes of the zooming range The zoom focal length ranges from 36 mm to 72 mm, the zoom ratio is $\gamma = 2$, the $F/\#$ is 4, the image height is 14 mm, and the focal lengths of the groups are -66 mm and 42 mm. The design is monochromatic, using glass LAK9 for both lenses, which are aspheric. Fourth-order coma and distortion remain uncorrected, in part because this simple zoom lens lacks symmetry about the stop aperture.

Figure 18.3 (right) shows a two-group zoom where two more lenses, one aspheric, have been added to the variator to help correct for coma and mitigate distortion aberration. The compensator is the negative single lens in front. This zoom lens is now monochromatically corrected for the primary aberrations at zoom range extremes, except distortion, which is about -5% in the wide angle position. The focal lengths of the groups are -67 mm and 37 mm.

Figure 18.4 shows the zoom lens optimized with real rays and achromatized in part by converting the central positive lens into a doublet. The prescription is given in Table 18.1. Only one aspheric surface was used in the zoom lens, and

Table 18.1 *Zoom lens prescription, f' = 36–72, FOV = ±30° to ±15°, F/4,
Conic constant surface 3, K = −1.7892*

Surface	Radius (mm)	Thickness (mm)	Glass
1	Plano	2.0	LAK34
2	48.7647	5.4–39.8	
3	47.0311	15.0	LAK33
4	−45.0391	4.0	LASF35
5	−88.8387	26.7700	
Stop		10.6901	
7	44.7092	5.0	SK4
8	−37.4045	7.0	
9	−23.3034	3.0	F14
10	73.9289	12.7–32.6	

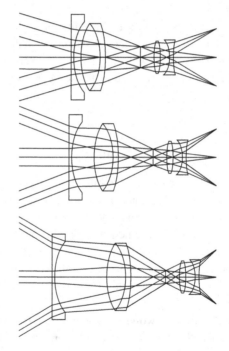

Figure 18.4 Zoom lens real ray optimized and achromatized for the *F* and *C*
wavelengths. f' = 36 mm to 72 mm, FOV = 15° to 30°, F/4.

is located in the doublet. The field of view was increased to ±30° (21 mm
image height) at the wide angle position. Figure 18.5 provides the RMS spot
size across the field of view for three zoom positions; other zoom positions
have comparable performance.

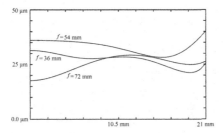

Figure 18.5 RMS spot size for three zoom positions: $f' = 36$ mm, $f' = 54$ mm, and $f' = 72$ mm.

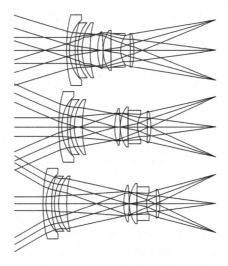

Figure 18.6 Two-group (NP) zoom lens by S. Sato, US Patent 4,792,215, Example 1. $f' = 36$–68.5 mm, FOV $= \pm32.2°$ to $\pm17.3°$, $F/4.1$.

By adding lens complexity, different zoom forms become possible; some with substantial improvement in performance. Figure 18.6 shows a two-group (NP) zoom lens from the patent literature. The back focal distance allows for a folding mirror for a single lens reflex (SLR) camera.

18.3 Three-Group Zoom

Zoom lenses with three groups have more design flexibility than two-group zoom lenses, and the zooming range can be larger. With three groups, several arrangements are possible. Two of them are the PNP and the NPN shown in Figure 18.7 and arranged as afocal systems with varying magnification; these

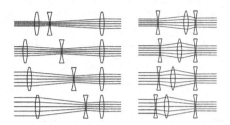

Figure 18.7 Donders type zoom telescopes; left, PNP; right, NPN.

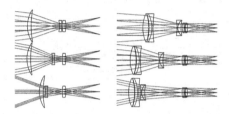

Figure 18.8 Left, Three-group zoom lens using individual aspheric lenses for each group. Spherical aberration, coma, and astigmatism have been corrected at the extremes of the zoom range, $\gamma = 3$. Right, Achromatic doublet and triplet lenses with spherical surfaces form the groups of a three-group zoom lens, $\gamma = 3$.

are known as the Donders type telescopes. A prime lens can be added at the rear end of the afocal zoom to form an image. The prime lens is also known as a relay lens, as it changes the location of the image.

Using individual lenses for each group, the three-group zoom lens system has enough degrees of freedom to correct for fourth-order spherical aberration, coma, and astigmatism at the extremes of the zoom range. In addition, one lens can be stationary and the zoom can be afocal, provide negative power, or provide positive power, as shown in Figure 18.8 (left). To achieve this correction, the degrees of freedom are the individual lenses optical power, the lens asphericity, and the lens air spaces. The use of an aspheric surface in each lens is temporary. Once the aberrations are corrected at two zoom positions, each of the lenses is split into a doublet or a triplet and, by bending the lenses, the aspheric surface is replaced with a spherical surface. Each group is also achromatized individually, as shown in Figure 18.8 (right).

Since the starting zoom lens is corrected for primary aberrations, an improved solution with real ray optimization often becomes feasible. Petzval field curvature is initially not corrected, but it can be corrected or balanced during real ray optimization by adding complexity to the stationary third group. Distortion aberration is corrected for an intermediate zoom position, also adding complexity in the third group, or by making the overall zoom lens more symmetrical about the aperture stop. However, the location of the stop affects lens diameter and light vignetting. Depending on the overall zoom lens, configuration distortion aberration might be tolerable at the zoom range extremes.

There is no need to make fourth-order spherical aberration, coma, and astigmatism equal to zero in the initial zoom kernel layout. Instead, these aberrations can be made to have the same value at either extreme of the zoom range, and later corrected by adding complexity to a stationary lens group. A zoom lens kernel can provide uniform aberration across the zooming range. This uniform aberration can often be corrected in a stationary lens group. As each lens group is further split, more degrees of freedom are gained to control aberration at the expense of lens complexity. Then, as each group becomes thicker, lens overlapping must be avoided.

The position of the compensator across the zooming range is known as the cam curve. Some attention must be given to how the moving groups change position; steep changes in the cam curves can be problematic for the zooming mechanism and for the user.

18.4 Four-Group Zoom

Further design flexibility is gained by adding a fourth lens group. There are many combinations for the sign of optical power and moving groups that can result from having four groups. For example, PNNP, PNPP, NPNP, NPPN. An additional group can be used to improve image quality, to extend the zooming range, or to meet additional requirements such as focusing. The zoom kernel typically works for one object position, and the image must be maintained in focus whenever the object distance changes. To compensate for this change, a first lens group, or focusing group, can help to maintain the conjugate distance at which the kernel groups work. However, focusing lens groups often change the field of view of a lens, resulting in an unnatural movement of objects in the field of view. This effect is known as breathing, and it may be objectionable in some applications.

An example of a four-group zoom lens is shown in Figure 18.9, which is from US Patent application 20090086321 by Keiko Mizuguchi. It is a PNPP zoom lens where the first and fourth groups do not move along the optical axis. The second and third groups form the zoom kernel. Seven spherical lenses make up the variator and compensator. The compensator also provides the

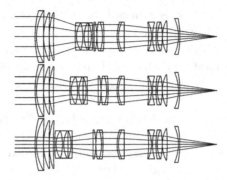

Figure 18.9 Four-group zoom lens from US Patent application 20090086321 by Keiko Mizuguchi. The first, second, and third groups form a Donders type telescope.

Figure 18.10 Forward and reverse ray tracing method to cancel aberrations from the intermediate zoom position. The aberrations from the kernel that depend on zoom position become obvious at the extreme positions of the zoom.

focusing function. The rear fourth group includes at least one lens that moves perpendicularly to the optical axis to compensate for vibration. The fourth group has enough complexity to control the aberrations that are independent of zooming. In addition, vignetting is used to control the overall lens diameter and aberration.

18.5 Zoom Lens Kernel

As discussed above, the use of aberration coefficients is helpful to analyze and design a zoom lens kernel. Another way to analyze, or optimize, a zoom lens kernel is by using reverse ray tracing. A given zoom lens is reversed and added to the rear end of the original zoom lens, as shown in Figure 18.10. A flat mirror is added at the original lens focal plane. The reversed lens has only one zoom position, say an intermediate position in the zooming range. Light rays

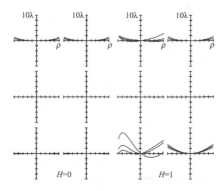

Figure 18.11 Wave fans for three zoom positions and three wavelengths. Top row, residual kernel aberrations at one zoom range extreme. Middle row, perfect cancelation of kernel aberrations. Bottom row, residual kernel aberrations at the other zoom range extreme.

will travel in the original zoom lens, and retrace their path in the reversed lens, to yield exactly zero aberration for the intermediate zoom position. At other zoom positions, the aberration is the difference of aberration of the zoom position in question and the intermediate zoom position. This difference in aberration must then be corrected. Light is collimated after it goes forward and backward through the zoom lens. The aberration analysis can be done in afocal mode. For this method to work it is required that the exit pupil remains stationary across the zooming range. Figure 18.11 shows zooming dependent aberrations, $W_K(\vec{H}\vec{\rho}\ \gamma)$, for three positions of a zoom lens using the reverse tracing method.

18.6 Aberration Considerations

There are several aberration considerations in the design of a zoom lens. According to stop shifting formulae, the chromatic change of magnification upon stop shifting, $\bar{S}$, depends on the chromatic change of focus,

$$\delta_\lambda W_{111}^* = \partial_\lambda W_{111} + 2\bar{S}\partial_\lambda W_{020}. \tag{18.3}$$

Because in a zoom lens the pupils may shift with respect to each group, then any group residual primary or secondary chromatic change of focus will generate chromatic change of magnification. This calls for at least making each group independently corrected for chromatic aberration.

Further, the chromatic change of magnification upon object shifting, S, depends on the chromatic change of focus of the pupil,

$$\partial_\lambda W^*_{111} = \partial_\lambda W_{111} + 2S\partial_\lambda \bar{W}_{020}. \tag{18.4}$$

As the object position for each zoom group can change according to zoom lens position, chromatic change of magnification can be generated.

The chromatic change of focus upon object shifting also depends on image and pupil chromatic aberrations,

$$\partial_\lambda W^*_{020} = \partial_\lambda W_{020} + \frac{1}{2}(\partial_\lambda W_{111} + \partial_\lambda \bar{W}_{111})S + S^2 \partial_\lambda \bar{W}_{020}. \tag{18.5}$$

Thus, it is necessary in the design of a zoom lens to consider pupil aberrations. The change of astigmatism aberration upon object shifting is given by,

$$W^*_{222} = W_{222} + \left(2W_{311} + \varkappa\Delta(\bar{u}^2)/2\right)S + 4\bar{W}_{040}S^2. \tag{18.6}$$

If a lens group is free from pupil spherical aberration, $\bar{W}_{040} = 0$, is corrected for image distortion, $W_{311} = 0$, and does not deviate the chief ray, $\Delta(\bar{u}^2) = 0$, then there is no change of astigmatism upon object shifting. Some zoom variators work near unit magnification, are symmetrical, and fulfil these conditions to some extent. Then such variators minimize the change of astigmatism aberration, and help to keep the zoom lens compact.

In combining a zoom lens with other lenses, one must keep in mind that the entrance and exit pupils of a zoom lens may change axial position, and then care must be paid to provide means for proper pupil matching.

Further Reading

Hopkins, H. H. "A class of symmetrical systems of variable power," Chapter 2 in *Optical Instruments*, Proceedings of the London Conference, 1950 (London: Chapman and Hall, 1951), 17–32.

Kingslake, R. "Varifocal and zoom lenses," Chapter 11, in *A History of the Photographic Lens* (San Diego, CA: Academic Press Inc., 1989).

Mann, Allen. *Selected Papers on Zoom Lenses* (Bellingham, WA: SPIE Milestone Series; V. MS 85, 1993).

Yan, Yufeng, Sasián, José. "Photographic zoom fisheye lens design for DSLR cameras," *Optical Engineering*, 56(9) (2017), 095103.

Youngworth, Richard N., Betensky, Ellis I. "Fundamental considerations for zoom lens design," Proceedings of SPIE 8488, Zoom Lenses IV, 848806 (2012); doi: 10.1117/12.930618.

Appendix 1

Imaging Aberrations

$$
\begin{aligned}
W\left(\vec{H},\vec{\rho}\right) = {}& W_{000} + W_{200}\left(\vec{H}\cdot\vec{H}\right) + W_{111}\left(\vec{H}\cdot\vec{\rho}\right) + W_{020}\left(\vec{\rho}\cdot\vec{\rho}\right) \\
& + W_{040}\left(\vec{\rho}\cdot\vec{\rho}\right)^2 + W_{131}\left(\vec{H}\cdot\vec{\rho}\right)\left(\vec{\rho}\cdot\vec{\rho}\right) + W_{222}\left(\vec{H}\cdot\vec{\rho}\right)^2 \quad\quad (A1.1) \\
& + W_{220}\left(\vec{H}\cdot\vec{H}\right)\left(\vec{\rho}\cdot\vec{\rho}\right) + W_{311}\left(\vec{H}\cdot\vec{H}\right)\left(\vec{H}\cdot\vec{\rho}\right) + W_{400}\left(\vec{H}\cdot\vec{H}\right)^2
\end{aligned}
$$

Table A1.1 *Primary aberration coefficients in terms of Seidel sums*

Coefficient	Seidel sum
$W_{040} = \dfrac{1}{8}S_I$	$S_I = -\sum_{i=1}^{j}\left(A^2 y\Delta\left(\dfrac{u}{n}\right)\right)_i$
$W_{131} = \dfrac{1}{2}S_{II}$	$S_{II} = -\sum_{i=1}^{j}\left(A\bar{A}y\Delta\left(\dfrac{u}{n}\right)\right)_i$
$W_{222} = \dfrac{1}{2}S_{III}$	$S_{III} = -\sum_{i=1}^{j}\left(\bar{A}^2 y\Delta\left(\dfrac{u}{n}\right)\right)_i$
$W_{220} = \dfrac{1}{4}(S_{IV} + S_{III})$	$S_{IV} = -\mathcal{K}^2\sum_{i=1}^{j}P_i$
$W_{311} = \dfrac{1}{2}S_V$	$S_V = -\sum_{i=1}^{j}\left(\bar{A}\left[\bar{A}^2\Delta\left(\dfrac{1}{n^2}\right)y - (\mathcal{K} + \bar{A}y)\bar{y}P\right]\right)_i$
$\delta_\lambda W_{020} = \dfrac{1}{2}C_L$	$C_L = \sum_{i=1}^{j}\left(Ay\Delta\left(\dfrac{\delta n}{n}\right)\right)_i$
$\delta_\lambda W_{111} = C_T$	$C_T = \sum_{i=1}^{j}\left(\bar{A}y\Delta\left(\dfrac{\delta n}{n}\right)\right)_i$

Table A1.2 *Quantities derived from first-order marginal and chief ray data used in computing the aberration coefficients*

Refraction invariant marginal ray	Refraction invariant chief ray	Lagrange invariant	Surface curvature	Petzval sum term
$A = ni = nu + nyc$	$\bar{A} = n\bar{i} = n\bar{u} + n\bar{y}c$	$\mathcal{K} = n\bar{u}y - nu\bar{y}$ $= \bar{A}y - A\bar{y}$	$c = \dfrac{1}{r}$	$P = c\cdot\Delta\left(\dfrac{1}{n}\right)$

Table A1.3 *Contributions to Seidel sums from an aspheric surface*

$$\delta S_I = a$$

$$\delta S_{II} = \left(\frac{\bar{y}}{y}\right) a$$

$$\delta S_{III} = \left(\frac{\bar{y}}{y}\right)^2 a$$

$$\delta S_{IV} = 0$$

$$\delta S_V = \left(\frac{\bar{y}}{y}\right)^3 a$$

$$\delta C_L = 0$$

$$\delta C_T = 0$$

$$a = -\varepsilon^2 c^3 y^4 \Delta(n) + 8A_4 y^4 \Delta(n)$$

For a conic surface of eccentricity, ε, and an aspheric fourth-order coefficient, A_4

Table A1.4 *Stop-shifting formulas*

$$\bar{S} = \frac{\bar{u}_{new} - \bar{u}_{old}}{u} = \frac{\bar{y}_{new} - \bar{y}_{old}}{y} = \frac{\bar{A}_{new} - \bar{A}_{old}}{A}$$

$$S_I^* = S_I$$

$$S_{II}^* = S_{II} + \bar{S}S_I$$

$$S_{III}^* = S_{III} + 2\cdot\bar{S}S_{II} + \bar{S}^2 S_I$$

$$S_{IV}^* = S_{IV}$$

$$S_V^* = S_V + \bar{S}(S_{IV} + 3\cdot S_{III}) + 3\cdot\bar{S}^2 S_{II} + \bar{S}^3 S_I$$

$$C_L^* = C_L$$

$$C_T^* = C_T + \bar{S}C_L$$

Table A1.5 *Object-shifting equations*

$$S = \frac{u^* - u}{\bar{u}} = \frac{y^* - y}{\bar{y}} = \frac{A^* - A}{\bar{A}}$$

$$W_{040}^* = W_{040} + \left(W_{131} + \frac{1}{8}\mathcal{K}\Delta(u^2)\right)S + \left(\frac{3}{2}W_{222} + \frac{3}{8}\mathcal{K}\Delta(u\bar{u}) + W_{220P}\right)S^2$$
$$+ \left(W_{311} + \frac{3}{8}\mathcal{K}\Delta(\bar{u}^2)\right)S^3 + \bar{W}_{040}S^4$$

$$W_{131}^* = W_{131} + \left(3W_{222} + \frac{1}{2}\mathcal{K}\Delta(u\bar{u}) + 2W_{220P}\right)S$$
$$+ (3W_{311} + \mathcal{K}\Delta(\bar{u}^2))S^2 + 4\bar{W}_{040}S^3$$

$$W_{220P}^* = W_{220P}$$

$$W_{222}^* = W_{222} + (2W_{311} + \mathcal{K}\Delta(\bar{u}^2)/2)S + 4\bar{W}_{040}S^2$$

$$W_{311}^* = W_{311} + 4\bar{W}_{040}S$$

$$\partial_\lambda W_{111}^* = \partial_\lambda W_{111} + 2\partial_\lambda \bar{W}_{020}S$$

$$\partial_\lambda W_{020}^* = \partial_\lambda W_{020} + \frac{1}{2}(\partial_\lambda W_{111} + \partial_\lambda \bar{W}_{111})S + \partial_\lambda \bar{W}_{020}S^2$$

Table A1.6 *Extrinsic coefficients from combining systems* A *and* B

$$W_{060E} = -\tfrac{1}{K}\left(4W_{040}^{A}\,\bar{W}_{311}^{B}\right)$$

$$W_{331E} = -\tfrac{1}{K}\begin{pmatrix} 5W_{131}^{A}\,\bar{W}_{131}^{B} + 4W_{220}^{A}\,\bar{W}_{220}^{B} \\ +4W_{220}^{A}\,\bar{W}_{222}^{B} + 4W_{222}^{A}\,\bar{W}_{220}^{B} \\ +\tfrac{311A}{W}\,\bar{W}_{311}^{B} + 16W_{040}^{A}\,\bar{W}_{040}^{B} \end{pmatrix}$$

$$W_{151E} = -\tfrac{1}{K}\begin{pmatrix} 3W_{131}^{A}\,\bar{W}_{311}^{B} + 8W_{040}^{A}\,\bar{W}_{220}^{B} \\ +8W_{040}^{A}\,\bar{W}_{222}^{B} \end{pmatrix}$$

$$W_{422E} = -\tfrac{1}{K}\begin{pmatrix} 2W_{311}^{A}\,\bar{W}_{222}^{B} + 4W_{220}^{A}\,\bar{W}_{131}^{B} \\ +6W_{222}^{A}\,\bar{W}_{131}^{B} + 8W_{131}^{A}\,\bar{W}_{040}^{B} \end{pmatrix}$$

$$W_{242E} = -\tfrac{1}{K}\begin{pmatrix} 2W_{222}^{A}\,\bar{W}_{311}^{B} + 4W_{131}^{A}\,\bar{W}_{220}^{B} \\ +6W_{131}^{A}\,\bar{W}_{222}^{B} + 8W_{040}^{A}\,\bar{W}_{131}^{B} \end{pmatrix}$$

$$W_{420E} = -\tfrac{1}{K}\begin{pmatrix} 2W_{220}^{A}\,\bar{W}_{131}^{B} + 2W_{311}^{A}\,\bar{W}_{220}^{B} \\ +4W_{131}^{A}\,\bar{W}_{040}^{B} \end{pmatrix}$$

$$W_{333E} = -\tfrac{1}{K}\left(4W_{131}^{A}\,\bar{W}_{131}^{B} + 4W_{222}^{A}\,\bar{W}_{222}^{B}\right)$$

$$W_{511E} = -\tfrac{1}{K}\begin{pmatrix} 3W_{311}^{A}\,\bar{W}_{131}^{B} + 8W_{220}^{A}\,\bar{W}_{040}^{B} \\ +8W_{222}^{A}\,\bar{W}_{040}^{B} \end{pmatrix}$$

$$W_{240E} = -\tfrac{1}{K}\begin{pmatrix} 2W_{131}^{A}\,\bar{W}_{220}^{B} + 2W_{220}^{A}\,\bar{W}_{311}^{B} \\ +4W_{040}^{A}\,\bar{W}_{131}^{B} \end{pmatrix}$$

$$W_{600E} = -\tfrac{1}{K}\left(4W_{311}^{A}\,\bar{W}_{040}^{B}\right)$$

Table A1.7 *Seidel sums of a parallel glass plate in air of index* n *and thickness* t

$$S_{I} = -\frac{n^2 - 1}{n^3}u^4 t$$

$$S_{II} = -\frac{n^2 - 1}{n^3}u^3\bar{u}t$$

$$S_{III} = -\frac{n^2 - 1}{n^3}u^2\bar{u}^2 t$$

$$S_{IV} = 0$$

$$S_{V} = -\frac{n^2 - 1}{n^3}u\bar{u}^3 t$$

$$C_{L} = -\frac{n - 1}{n^2 v}u^2 t$$

$$C_{T} = -\frac{n - 1}{n^2 v}u\bar{u}t$$

Appendix 2

Pupil Aberrations

$$\bar{W}\left(\vec{H}, \vec{\rho}\right) = \bar{W}_{000} + \bar{W}_{200}\left(\vec{\rho} \cdot \vec{\rho}\right) + \bar{W}_{111}\left(\vec{H} \cdot \vec{\rho}\right) + \bar{W}_{020}\left(\vec{H} \cdot \vec{H}\right)$$
$$+ \bar{W}_{040}\left(\vec{H} \cdot \vec{H}\right)^2 + \bar{W}_{131}\left(\vec{H} \cdot \vec{H}\right)\left(\vec{H} \cdot \vec{\rho}\right) + \bar{W}_{222}\left(\vec{H} \cdot \vec{\rho}\right)^2$$
$$+ \bar{W}_{220}\left(\vec{H} \cdot \vec{H}\right)\left(\vec{\rho} \cdot \vec{\rho}\right) + \bar{W}_{311}\left(\vec{\rho} \cdot \vec{\rho}\right)\left(\vec{H} \cdot \vec{\rho}\right) + \bar{W}_{400}\left(\vec{\rho} \cdot \vec{\rho}\right)^2$$

$$(A2.1)$$

Table A2.1 *Identities between pupil and image aberration coefficients*

$\bar{W}_{040} = W_{400}$
$\bar{W}_{131} = W_{311} + \dfrac{1}{2}\mathcal{K}\cdot\Delta\{\bar{u}^2\}$
$\bar{W}_{222} = W_{222} + \dfrac{1}{2}\mathcal{K}\cdot\Delta\{u\bar{u}\}$
$\bar{W}_{220} = W_{220} + \dfrac{1}{4}\mathcal{K}\cdot\Delta\{u\bar{u}\}$
$\bar{W}_{311} = W_{131} + \dfrac{1}{2}\mathcal{K}\cdot\Delta\{u^2\}$
$\bar{W}_{400} = W_{040}$

Table A2.2 *Pupil chromatic coefficients*

$\delta_\lambda \bar{W}_{020} = \dfrac{1}{2}\bar{C}_L$	$\bar{C}_L = \sum\limits_{i=1}^{j}\left(\bar{A}\bar{y}\Delta\left(\dfrac{\delta n}{n}\right)\right)_i$
$\delta_\lambda \bar{W}_{111} = \bar{C}_T$	$\bar{C}_T = \sum\limits_{i=1}^{j}\left(A\bar{y}\Delta\left(\dfrac{\delta n}{n}\right)\right)_i$
$\bar{C}_L^* = \bar{C}_L$	$\bar{C}_T^* = \bar{C}_T + S\bar{C}_L$

Appendix 3

Structural Aberration Coefficients

Table A3.1 *Seidel sums in terms of structural aberration coefficients. Pupils located at principal planes*

$$S_I = \frac{1}{4} y_P^4 \Phi^3 \sigma_I$$

$$S_{II} = \frac{1}{2} \mathcal{K} y_P^2 \Phi^2 \sigma_{II}$$

$$S_{III} = \mathcal{K}^2 \Phi \sigma_{III}$$

$$S_{IV} = \mathcal{K}^2 \Phi \sigma_{IV}$$

$$S_V = \frac{2 \mathcal{K}^3 \sigma_V}{y_P^2}$$

$$C_L = y_P^2 \Phi \sigma_L$$

$$C_T = 2 \mathcal{K} \sigma_T$$

Table A3.2 *Stop-shifting from principal planes*

$$\sigma_I^* = \sigma_I$$
$$\sigma_{II}^* = \sigma_{II} + \bar{S}_\sigma \sigma_I$$
$$\sigma_{III}^* = \sigma_{III} + 2\bar{S}_\sigma \sigma_{II} + \bar{S}_\sigma^2 \sigma_I$$
$$\sigma_{IV}^* = \sigma_{IV}$$
$$\sigma_V^* = \sigma_V + \bar{S}_\sigma(\sigma_{IV} + 3\sigma_{III}) + 3\bar{S}_\sigma^2 \sigma_{II} + \bar{S}_\sigma^3 \sigma_I$$
$$\sigma_L^* = \sigma_L$$
$$\sigma_T^* = \sigma_T + \bar{S}_\sigma \sigma_L$$
$$\bar{S}_\sigma = \frac{y_P \bar{y}_P \Phi}{2\mathcal{K}}$$
$$\Delta\bar{S}_\sigma = \frac{y_P \Delta\bar{y}_P \Phi}{2\mathcal{K}} = \frac{y_P^2 \Phi}{2\mathcal{K}} \bar{S}$$
$$\bar{S}_\sigma = \frac{y_P \bar{y}_P \Phi}{2\mathcal{K}} = \frac{\Phi \cdot \bar{s}}{(Y - 1)\cdot\Phi\cdot\bar{s} - 2n} = \frac{\Phi \cdot \bar{s}'}{(Y + 1)\cdot\Phi\cdot\bar{s}' - 2n'}$$

$\bar{s}$ is the distance from the front principal point to the entrance pupil. $\bar{s}'$ is the distance from the rear principal point to the exit pupil.

y_P is the marginal ray height at the principal planes. $\bar{y}_P$ is the chief ray height at the principal planes. Φ is the lens system optical power.

Table A3.3 *Field curve vertex curvature in terms of structural coefficients*

$$C_{Petzval} = -n'\Phi\cdot\sigma_{IV}$$
$$C_{Sagittal} = -n'\Phi\cdot(\sigma_{IV} + \sigma_{III})$$
$$C_{Medial} = -n'\Phi\cdot(\sigma_{IV} + 2\sigma_{III})$$
$$C_{Tangential} = -n'\Phi\cdot(\sigma_{IV} + 3\sigma_{III})$$

Table A3.4 *Structural aberration coefficients of a reflecting surface in air*

Stop at surface	With stop shift
$\sigma_I = Y^2 + K$	$\sigma_I = Y^2 + K$
$\sigma_{II} = -Y$	$\sigma_{II} = -Y(1 - \bar{S}_\sigma Y) + \bar{S}_\sigma\cdot K$
$\sigma_{III} = 1$	$\sigma_{III} = (1 - \bar{S}_\sigma Y)^2 + \bar{S}_\sigma^2\cdot K$
$\sigma_{IV} = -1$	$\sigma_{IV} = -1$
$\sigma_V = 0$	$\sigma_V = \bar{S}_\sigma\cdot(1 - \bar{S}_\sigma Y)\cdot(2 - \bar{S}_\sigma Y) + \bar{S}_\sigma^3\cdot K$

K is the surface conic constant, $Y = \dfrac{1 + m}{1 - m}$, $\bar{S}_\sigma = \dfrac{y_P \bar{y}_P \Phi}{2\mathcal{K}}$

Table A3.5 *First-order identities of a thin lens*

$$\Phi = (n-1) \cdot (c_1 - c_2) = (n-1) \cdot \left(\frac{1}{r_1} - \frac{1}{r_2} \right)$$

$$X = \frac{c_1 + c_2}{c_1 - c_2} = -\frac{r_1 + r_2}{r_1 - r_2}$$

$$Y = \frac{1+m}{1-m}$$

$$c_1 = \frac{1}{2} \frac{\Phi}{n-1} (X+1)$$

$$c_2 = \frac{1}{2} \frac{\Phi}{n-1} (X-1)$$

$$\omega = nu = -\frac{1}{2}(Y-1)(\Phi \cdot y_P)$$

$$\omega' = n'u' = -\frac{1}{2}(Y+1)(\Phi \cdot y_P)$$

Table A3.6 *Structural aberration coefficients of a thin lens in air (Stop at lens)*

$\sigma_I = AX^2 - BXY + CY^2 + D$	$A = \dfrac{n+2}{n(n-1)^2}$
$\sigma_{II} = EX - FY$	$B = \dfrac{4(n+1)}{n(n-1)}$
$\sigma_{III} = 1$	$C = \dfrac{3n+2}{n}$
$\sigma_{IV} = \dfrac{1}{n}$	$D = \dfrac{n^2}{(n-1)^2}$
$\sigma_V = 0$	$E = \dfrac{n+1}{n(n-1)}$
$\sigma_L = \dfrac{1}{v}$	$F = \dfrac{2n+1}{n}$
$\sigma_T = 0$	$v = \dfrac{n_F - n_C}{n_d - 1}$

Table A3.7 *Structural coefficients of a system of k components*

$$\sigma_I = \sum_{i=1}^{k} \left(\frac{\Phi_k}{\Phi}\right)^3 \left(\frac{y_{P,k}}{y_P}\right)^4 \sigma_{I,k}$$

$$\sigma_{II} = \sum_{i=1}^{k} \left(\frac{\Phi_k}{\Phi}\right)^2 \left(\frac{y_{P,k}}{y_P}\right)^2 (\sigma_{II,k} + \bar{S}_k \sigma_{I,k})$$

$$\sigma_{III} = \sum_{i=1}^{k} \left(\frac{\Phi_k}{\Phi}\right) \left(\sigma_{III,k} + 2\bar{S}_k \sigma_{II,k} + \bar{S}_k^2 \sigma_{I,k}\right)$$

$$\sigma_{IV} = \sum_{i=1}^{k} \left(\frac{\Phi_k}{\Phi}\right) \sigma_{IV,k}$$

$$\sigma_V = \sum_{i=1}^{k} \left(\frac{y_P}{y_{P,k}}\right)^2 \left(\sigma_{V,k} + \bar{S}_k(\sigma_{IV,k} + 3\sigma_{III,k}) + 3\bar{S}_k^2 \sigma_{II,k} + \bar{S}_k^3 \sigma_{I,k}\right)$$

$$\sigma_L = \sum_{i=1}^{k} \left(\frac{\Phi_k}{\Phi}\right) \left(\frac{y_{P,k}}{y_P}\right)^2 \sigma_{L,k}$$

$$\sigma_T = \sum_{i=1}^{k} (\sigma_{T,k} + \bar{S}_k \sigma_{L,k})$$

$$\bar{S}_k = \frac{\Phi_k \cdot y_{P,k} \cdot \bar{y}_{P,k}}{2\text{Ж}}$$

Appendix 4

Primary Aberrations of a Plane Symmetric System

To describe the aberration properties of a plane symmetric system, an aberration function must be constructed. We establish the unit vector, $\vec{i}$, in the field of view to define the direction of plane of symmetry. Since the aberration function is a scalar, it must depend on the dot products of the field vector, the aperture vector, and the symmetry vector, $\vec{i}$. The aberration function for a plane symmetric system can be written as,

$$W\left(\vec{i},\vec{H},\vec{\rho}\right)= \sum_{\substack{k,m,n,p,q}} W_{\substack{2k+n+p,\\2m+n+q,\\n,p,q}} \left(\vec{H}\cdot\vec{H}\right)^{k}\left(\vec{\rho}\cdot\vec{\rho}\right)^{m}\left(\vec{H}\cdot\vec{\rho}\right)^{n}\left(\vec{i}\cdot\vec{H}\right)^{p}\left(\vec{i}\cdot\vec{\rho}\right)^{q},$$

(A4.1)

where $W_{2k+n+p,\,2m+n+q,\,n,p,q}$ is the coefficient of a particular aberration form defined by the integers, k, m, n, p, and q. By setting the sum of the integers to 0, 1, 2, ..., groups of aberrations are defined as shown in Table A4.1.

A study of Table A4.1 provides useful insights. The aberration terms are divided in groups and in turn in subgroups according to symmetry characteristics. Thus, the third group contains the primary aberrations of axially symmetric systems as a subgroup, contains the aberrations of double plane symmetric systems as a subgroup, and a subgroup of aberrations for plane symmetric systems that are not axially or double plane symmetric. Thus, the aberration properties of a plane symmetric system can be thought of as the superposition of the properties of axial, double plane, and plane symmetric systems. The correction of the aberrations of a given subgroup can be carried out using system properties according to subgroup symmetry.

Table A4.1 *Aberrations of a plane symmetric system*

First group	
W_{00000}	Piston
Second group	
$W_{01001}\left(\vec{i}\cdot\vec{\rho}\right)$	Field displacement
$W_{10010}\left(\vec{i}\cdot\vec{H}\right)$	Linear Piston
$W_{02000}\left(\vec{\rho}\cdot\vec{\rho}\right)$	Defocus
$W_{11100}\left(\vec{H}\cdot\vec{\rho}\right)$	Magnification
$W_{20000}\left(\vec{H}\cdot\vec{H}\right)$	Quadratic Piston
Third group	
$W_{02002}\left(\vec{i}\cdot\vec{\rho}\right)^{2}$	Uniform astigmatism
$W_{11011}\left(\vec{i}\cdot\vec{H}\right)\left(\vec{i}\cdot\vec{\rho}\right)$	Anamorphic distortion
$W_{20020}\left(\vec{i}\cdot\vec{H}\right)^{2}$	Quadratic piston
$W_{03001}\left(\vec{i}\cdot\vec{\rho}\right)\left(\vec{\rho}\cdot\vec{\rho}\right)$	Uniform coma
$W_{12101}\left(\vec{i}\cdot\vec{\rho}\right)\left(\vec{H}\cdot\vec{\rho}\right)$	Linear astigmatism
$W_{12010}\left(\vec{i}\cdot\vec{H}\right)\left(\vec{\rho}\cdot\vec{\rho}\right)$	Field tilt
$W_{21001}\left(\vec{i}\cdot\vec{\rho}\right)\left(\vec{H}\cdot\vec{H}\right)$	Quadratic distortion
$W_{21110}\left(\vec{i}\cdot\vec{H}\right)\left(\vec{H}\cdot\vec{\rho}\right)$	Quadratic distortion
$W_{30010}\left(\vec{i}\cdot\vec{H}\right)\left(\vec{H}\cdot\vec{H}\right)$	Cubic piston
$W_{04000}\left(\vec{\rho}\cdot\vec{\rho}\right)^{2}$	Spherical aberration
$W_{13100}\left(\vec{H}\cdot\vec{\rho}\right)\left(\vec{\rho}\cdot\vec{\rho}\right)$	Linear coma
$W_{22200}\left(\vec{H}\cdot\vec{\rho}\right)^{2}$	Quadratic astigmatism
$W_{22000}\left(\vec{H}\cdot\vec{H}\right)\left(\vec{\rho}\cdot\vec{\rho}\right)$	Field curvature
$W_{31100}\left(\vec{H}\cdot\vec{H}\right)\left(\vec{H}\cdot\vec{\rho}\right)$	Cubic distortion
$W_{40000}\left(\vec{H}\cdot\vec{H}\right)^{2}$	Quartic piston

Further Reading

Moore, Lori B., Hvisc, Anastacia M., Sasián, José. "Aberration fields of a combination of plane symmetric systems," *Optics Express*, 16 (2008), 15655–70.

Reshidko Dmitry, Sasián, José. "A method for the design of unsymmetrical optical systems using freeform surfaces," Proceedings of SPIE 10590, International Optical Design Conference 2017, 105900V (2017), doi: 10.1117/12.2285134.

Sasián, José. "Method of confocal mirror design," *Optical Engineering*, 58(1) (2019), 015101.

Sasián, José M. "Review of methods for the design of unsymmetrical optical systems," Proceedings of SPIE 1396, Applications of Optical Engineering: Proceedings of OE/Midwest '90 (1991), doi: 10.1117/12.25846.

Sasián, José M. "How to approach the design of a bilateral symmetric optical system," *Optical Engineering*, 33(6) (1994), 2045–61.

Sasián, José M. "Double-curvature surfaces in mirror system design," *Optical Engineering*, 36(1) (1997), 183–88.

Appendix 5

Sine Condition

The sine condition is theoretically an important result, as an aplanatic lens system must satisfy it. In the derivation that follows we assume that there is no spherical aberration in the lens system. Consider Figure A5.1, which shows the optical axis, the object and image planes, the on-axis conjugated points, O and O', the surfaces of unit magnification that near the optical axis region approach the principal planes, and the points, P and P', that are conjugated to second-order.

The optical path length along any ray from the on-axis object point, O to the image point O', is constant and is denoted by $l_{on-axis}$. We are interested in determining the optical path for a meridional ray from an off-axis point near the optical axis, as specified by heights, h and h'. With respect to the optical path, $l_{on-axis}$, and noting that points P and P' are conjugated, the optical path of an off-axis ray, $l_{off-axis}$, is given to first-order in h by,

$$
\begin{aligned}
l_{off-axis} &= l_{on-axis} - n \sin (U) h + n' \sin (U') h' \\
&= l_{on-axis} + nh \left(\frac{n'h'}{nh} \sin (U') - \sin (U) \right),
\end{aligned}
\tag{A5.1}
$$

where U and U' are the angles of a real marginal real ray with the optical axis in object and image spaces, respectively. For small object heights, h, the ratio h/h' becomes the transverse magnification $n'u'/nu$, fourth and higher order path differences between points P and P' become insignificant, and to have the path, $l_{off-axis}$, independent of the object height, h, we must satisfy,

$$
\frac{u'}{u} = \frac{\sin (U')}{\sin (U)}.
\tag{A5.2}
$$

This relationship is known as the sine condition and states that, in the absence of spherical aberration, there are no linear phase changes as a function of the

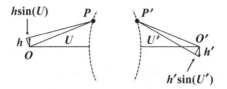

Figure A5.1 Geometry for deriving the sine condition. On-axis points, O and O', are conjugate and free from spherical aberration.

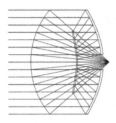

Figure A5.2 Geometry illustrating the equivalent refracting surface of an aplanatic Cassegrain type telescope as a sphere of radius f. The surface defined by the intersection of the incoming parallel rays in object space, with the outgoing rays in image space (as these rays are extended) is the principal surface; in this case the cross-section is circular.

field of view. In this case, the first-order magnification, $n'u'/nu$, is equal to the marginal real ray magnification, as defined by $n' \sin(U')/n \sin(U)$.

In deriving the sine condition we have made a first-order approximation with respect to the field of view and, therefore, all aberration coefficients of the form $W_{1,2n+1,1}$ vanish, with n an integer. Effectively, we can write,

$$\left(\sum_{n=1}^{\infty} W_{1,2n+1,1} \right) = l_{off-axis} - l_{on-axis} = n' \sin(U')h' - n \sin(U)h. \quad \text{(A5.3)}$$

Spherical aberration is independent of the field of view and, if it is added to an aplanatic system, say at the exit pupil, then the sine condition would no longer hold, but the system would still be free of coma aberration. Fundamentally, and due to conservation of throughput, the absence of linear phase variations with respect to the field of view, coma aberration in the present case, requires a specific pupil distortion that, to fourth-order, is given by,

$$\bar{W}_{311} = \frac{\mathcal{K}}{2} \Delta(u^2). \quad \text{(A5.4)}$$

For an object at infinity, the ratio $\sin(U)/u$ becomes unity. Therefore, the sine condition becomes,

$$\frac{\sin\left(U'\right)}{u'} = 1. \tag{A5.5}$$

This equation is satisfied if the principal surface is a sphere of radius equal to the lens system focal length, as shown in Figure A5.2 for the Ritchey-Chretien telescope.

Further Reading

Hopkins, H. H. "The sine condition and the Herschel's condition," Chapter III, in *Wave Theory of Aberrations* (Oxford: Oxford University Press, 1950).

Glossary

Abbe number A number that characterizes light dispersion by a material and is given by $v = \dfrac{n_d - 1}{n_F - n_C}$.

Aberration A departure from ideal imaging behavior. The classical imaging aberrations are spherical, coma, astigmatism, field curvature, distortion, chromatic change of focus, and chromatic change of magnification.

Aberration function A scalar mathematical function that provides the optical path length of a given ray that is specified by the field vector and the aperture vector. The aberration function is expressed as a series of terms of increasing algebraic order, each representing an optical aberration.

Achromatic Freedom from chromatic aberration for two wavelengths; typically those corresponding to the F and C lines.

Afocal The focal lengths are not defined. The optical power of an afocal system is zero.

Air space The axial separation between two contiguous lens or mirror elements.

Aligned An optical system where errors in the position of the lens elements do not significantly degrade the intended system performance.

Anamorphic Having different transverse magnification in two mutually perpendicular directions.

Anastigmatic In the context of a flat field lens, anastigmatic refers to the freedom from spherical aberration, coma, and astigmatism aberrations. In particular, there is no astigmatism aberration for an off-axis field point.

Aperture stop A circular aperture that limits in extent the size of the on-axis light beam.

Aperture vector A normalized vector, $\vec{p}$, that specifies a ray. It usually lies on the exit pupil plane of a lens system.

Aplanatic Freedom from spherical aberration and coma aberration.

Apochromatic Freedom from chromatic aberration for three wavelengths.

Aspheric A non-spherical in shape surface.

Aspheric plate A nearly parallel glass plate that has at least one aspheric surface, such as in the Schmidt camera.

Axial symmetry Having an axis of rotational symmetry.

Back focal length (BFL) The distance from the vertex of the last surface in a lens system to the image. Also known as back focal distance.

Bending Changing the surface curvatures of a lens element without changing its optical power.

Birefringence Double refraction in which two rays, two wavefronts, and two images are produced by a birefringent material.

Bore sight error The line of sight is in error; or, equivalently, the center of the field of view, is not in its intended position. An alignment error.

Cam curve The curve defined by the movement of a compensator group in a zoom lens.

Cardinal points The focal points, the nodal points, and the principal points of a lens system.

Catadioptric Comprising light reflecting mirrors and light refracting lenses.

Catoptric Comprising only light reflecting mirrors.

Chief ray A meridional ray that originates from the edge of the field of view and that passes through the center of the aperture stop.

Clear aperture diameter The minimum diameter of a lens necessary to allow light to pass through the lens. The diameter of the lens that is optically useful for the lens to perform its function. See lens diameter.

Coddington equations Relate the distances of the astigmatic line segments to the principal curvatures at a point of a surface. The sagittal equation is

$$\frac{n'}{s'} - \frac{n}{s} = \frac{n' \cos(I') - n \cos(I)}{r_s},$$

and the tangential equation is

$$\frac{n' \cos^2(I')}{t'} - \frac{n \cos^2(I)}{t} = \frac{n' \cos(I') - n \cos(I)}{r_t}.$$

Collimated A beam of light rays that is parallel and lacks convergence or divergence. A lens system that has been adjusted to produce a beam of parallel rays, or that has been aligned.

Comachromatism The variation of coma aberration with the wavelength of light.

Compensator In a zoom lens the group that re-establish the image location. In lens tolerancing a lens element parameter that can be used to compensate for lens system performance degradation due to manufacturing and alignment errors in other lens elements.

Concave surface In the lens design art a concave surface refers to a surface made in a substrate. Assuming that the surface is spherical and that the opposing surface of the substrate is flat, a concave surface makes the substrate thicker at its edges than at its center.

Conicoid A surface of revolution produced by rotating a conic curve about its axis of symmetry; i.e. hyperboloid, ellipsoid, paraboloid, and spheroid.

Convex surface In the lens design art, a convex surface refers to a surface made in a substrate. Assuming that the surface is spherical and that the opposing surface of the substrate is flat, a convex surface makes the substrate thicker at its center than at its edges.

Curvature For a spherical surface the inverse of the radius of curvature.

Defocus A change of the axial position of an image or an observation plane. It relates to a quadratic wavefront deformation, W_{020}. Also known as focus error.

Depth of field Relates to the object space; it is the axial distance an object can be moved while maintaining a given image quality at the image plane.

Depth of focus Relates to the image space; it is the axial distance that a lens or a light sensor can be moved while maintaining a given image quality for a given object.

Diffraction limited The image quality is limited by the effects of light diffraction and not by geometrical aberrations. Often, if the RMS wavefront deformation is less than 0.07 λ, a system is considered diffraction limited. However, in some applications, like optical lithography or microscopy, the requirement on RMS wavefront error is more stringent.

Diopter The unit of optical power, expressed in inverse meters.

Dioptric Comprising only light refracting lenses.

Doublet lens A lens system, or lens group, having two lens elements in proximity or in contact.

Dummy surface A surface with the same material on either side and that is used to determine light beam size, to reference distances, and for other purposes. It has no ray deviation function.

Effective focal length (EFL) The inverse of the optical power of a lens system.

Entrance pupil The image of the aperture stop in object space.

Error function A function that conveys the optical performance of a lens system according to the application of the system. It is also known as the merit function.

Etendue or throughput A geometrical quantity that measures the capacity of a light source, optical system, or image sensor to deliver, transfer, or receive optical power.

Exit pupil The image of the aperture stop in image space.

Eyepiece In a visual instrument the lens, or lens group, closest to the observing eye, and that collimates light.

Field flattener A field lens that corrects for field curvature aberration.

Field lens A lens placed at or near an image.

Field of view (FOV) The observable scene of a lens system for which it was designed. For a lens that works at finite conjugates the FOV is specified by the object or image size, and giving the height, width, or both. When the object is at infinity the field of view is usually specified by the semi-angle subtended by the scene, or object, as seen from the entrance pupil, either horizontally, vertically, or both. It can also be specified by the height, width, or both, of the image.

Field stop An aperture that limits the field of view of a lens system.

Field vector A normalized vector, $\vec{H}$, that specifies a ray. It lies on the object plane of a lens system.

Figure The shape of an optical surface, or the shape error of a surface.

First-order ray A ray traced with the first-order refraction and transfer equations, $n'u' = nu - y\phi$ and $y' = y + u't$.

Fisheye lens A reverse telephoto lens that images a hemispherical field of view, i.e. the field of view is $\pm 90°$.

Flare Non-uniform light reaching the image plane such as a defocused ghost image of the aperture stop.

Focal length The distance from the rear nodal point to the rear focal point is the rear focal length, f'. The distance from the front nodal point to the front focal point is the front focal length f.

Focal point The conjugate to the point at infinity in image space is the rear focal point. The conjugate to the point at infinity in object space is the front focal point. First-order rays converge to a focal point, or appear to diverge from a focal point.

Focal ratio See relative aperture.

Freeform surface An aspheric optical surface that does not have more than one plane of mirror symmetry.

Ghost image A non-intended image formed near or at the nominal image plane of a lens by light that is Fresnel reflected by one or more surfaces in a lens system.

Glass code A way to specify a glass type by the index of refraction and the v-number. For BK7 glass, $n_d = 1.516$ and $v = 64.1$, the glass code is 516641.

Gradient index The transverse or longitudinal variation of the index of refraction in a lens.

Ideal image An image defined by an ideal imaging model. For example, by the central projection model, as specified by Gaussian or Newtonian imaging equations. This is equivalent to the imaging defined by first-order ray tracing.

IR Infrared radiation.

Iris diaphragm A variable aperture to control the amount of light through a lens. Often it is the lens aperture stop.

Lagrange invariant A first-order quantity, $\mathcal{K} = n\bar{u}y - nu\bar{y} = \bar{A}y - A\bar{y}$, built with data of the marginal and chief rays that does not depend on the plane where it is calculated. It determines the optical throughput of a lens system.

Lens A single element or a complete optical system.

Lens decenter An error in the transverse position of a lens.

Lens diameter The physical diameter of a circular lens. It is larger than the clear aperture diameter to allow for lens mounting.

Lens element A single lens with two optical surfaces.

Lens group In the context of zoom lenses, one or more consecutive lens elements with their respective distances fixed.

Lens hood A mechanical structure, or shade, added to the front of a lens to avoid unwanted light from a bright source entering the lens.

Lens kernel The variator and the compensator groups in a zoom lens.

Lens maker's formula Relates the lens surface radius of curvature and thickness to the lens focal length, $\frac{1}{f'} = (n-1)\left(\frac{1}{r_1} - \frac{1}{r_2} + \frac{t\,n-1}{n\,r_1 r_2}\right)$.

Lens splitting Dividing a lens into two lenses close together and usually each with half the optical power of the un-split lens.

Lens system One or more lens or mirror elements that perform an optical function; usually that of forming an image.

Lens tilt An angular error in the position of a lens.

Lens unit One or more lens elements that perform an optical function such as focusing, correcting field aberrations, changing the magnification, or correcting for line of sight error.

Light propagation In modeling a lens system it is customary that light rays propagate from left to right. When there is a single mirror, light propagates from right to left after being reflected by the mirror. A second flat mirror can be added in contact to a first mirror to maintain light rays propagating from left to right. This is called unfolding the ray path.

Magnifying power In a visual instrument the ratio of the apparent size of the image of an object seen through the instrument to that of the object seen by the unaided eye. In a telescope, this is given by the ratio of the entrance pupil diameter to the exit pupil diameter.

Mangin mirror or lens A lens with one of its optical surfaces silvered or aluminized to be reflective.

Marginal ray A meridional ray that originates from the object axial point and passes through the edge of the aperture stop.

Meniscus lens A lens with a concave surface and a convex surface.

Meridional plane A plane that contains the optical axis.

Negative lens A lens with negative optical power, as determined using the vertex surface curvatures. A thin negative lens with spherical surfaces is thicker at its edge than at its center.

Nodal points A pair of points in the optical axis such that a first-order ray passing by one, passes by the other, while maintaining the same slope with the optical axis. The nodal points are conjugated and are the centers of perspective.

Null corrector An auxiliary lens system used to test an aspheric optical surface by forming a point image.

Numerical aperture (NA) The numerical aperture is defined as $NA = n \sin(\theta)$, where n is the index of refraction and θ is the angle of the real marginal ray with the optical axis.

Objective lens In a lens system the lens or lens group that is closest to the object and that focuses light.

Optical axis Usually a straight line about which a lens system has rotational symmetry. In some other contexts a straight line about which an optical system has some degree of symmetry, or property, or that serves as a reference.

Optical image A representation of an object by light. The image formed by a lens system.

Optical path difference The difference in optical path length of a ray between the wavefront and the reference sphere.

Optical power The optical power is defined as $\phi = -\dfrac{n}{f} = \dfrac{n'}{f'}$. For a single surface, the optical power is $\phi = \dfrac{n' - n}{r}$, where r is the vertex radius of curvature. The optical power of a thick lens in air is $\phi = (n - 1)\left(\dfrac{1}{r_1} - \dfrac{1}{r_2} + \dfrac{t}{n}\dfrac{n - 1}{r_1 r_2}\right)$, where t is the lens thickness measured along the optical axis.

Optical speed Refers to the time necessary to make a light exposure on a photographic film which is proportional to square of the lens f-number. Also known as lens speed.

Optically conjugated Satisfying the Gaussian or Newtonian imaging equations.

Petzval sum Relates the vertex curvature of the object and image surfaces to the index of refraction and surface vertex radius of curvature,
$$\frac{1}{n_k'\rho_k'} - \frac{1}{n_1\rho_1} = -\sum_{i=1}^k \frac{n_i' - n_i}{n_i n_i' r_i}.$$

Petzval surface The imaging surface in the absence of astigmatism aberration.

Plane symmetry Having a plane of mirror symmetry such as a meridional plane that contains the optical axis.

Positive lens A lens with positive optical power, as determined using the vertex surface curvatures. A thin positive lens with spherical surfaces is thicker at its center than at its edges.

Prescription table A table that provides constructional data of a lens such as radii, thicknesses, glass type, aspheric coefficients, etc., and main specifications of the lens such as FOV, focal length, optical speed, etc. Unless otherwise specified, axial symmetry of the lens system is assumed.

Principal plane Conjugate planes in a lens system having unit magnification.

Principal points The on-axis points of the principal planes in a lens system.

Principal ray The ray of an off-axis beam that passes through the center of the aperture stop.

Principle of symmetry When a lens system has some degree of symmetry about the aperture stop, the odd aberrations cancel or tend to cancel.

Real image An image that can be cast on a screen. The imaging rays converge toward the image.

Real ray A ray traced using Snell's law and the actual shape and position of the surfaces of an optical system.

Reference sphere A sphere centered at an ideal image point that passes by the axial exit pupil point. It is used to determine the wavefront deformation.

Refractive power Same as optical power.

Relative aperture The ratio of the effective focal length, EFL, to the diameter of the entrance pupil, D_E. Also known as $F/\#$, FNO, F-number, focal ratio F. $F/\# = EFL/D_E$. For lens systems that work at finite conjugates, the effective relative aperture, sometimes referred to as the working $F/\#$, is given by $F/\# = (1-m)EFL/D_E$, where m is the transverse magnification.

Relative illumination The ratio of the off-axis light irradiance to the on-axis light irradiance at the image plane.

Rotational invariants The products $\vec{H} \cdot \vec{H}, \vec{H} \cdot \vec{\rho}$, and $\vec{\rho} \cdot \vec{\rho}$, which are invariant upon a rotation of the coordinate axes.

Sag Short for sagittal depth of a surface.

Sagittal In a direction perpendicular to the tangent line of a curve.

Sagittal plane A plane perpendicular to a meridional plane and that contains a chief or principal ray.

Seidel sums Five formulas to determine aberration coefficients in an axially symmetric system from data of the trace of first-order marginal and chief rays.

Skew ray A ray not contained in a meridional plane. Its ray tracing requires quantities in three dimensions.

Snell's law Relates the angle of incidence and refraction to the indices of refraction: $n' \sin (I') = n \sin (I)$. The angle of incidence and refraction are with respect to the surface normal line where a ray intersects the surface. The normal line, the incident ray, and the refracted ray are coplanar.

Spherochromatism The variation of spherical aberration with the wavelength of light.

Stigmatic That produces a point image.

Stop aperture Same as aperture stop.

Stop shifting Moving the stop aperture along the optical axis and changing its diameter to maintain the optical throughput of the lens system.

Stray light Un-wanted and usually non-uniform light reaching the image plane and decreasing the image contrast, or forming image artifacts.

Structural aberration coefficient A coefficient that describes aberration in a simple mathematical form by not depending on the field of view, optical speed, or optical power, but on the structure of the lens system.

Tangential plane A plane containing the optical axis; same as a meridional plane.

Telecentric The entrance or exit pupil is at infinity. This requires having the aperture stop located at one focal point. The first-order chief ray is parallel to the optical axis in object or image spaces.

Telephoto lens A lens system where the ratio of the total track length to the focal length, called the telephoto ratio, is smaller than one.

Thin lens A conceptual model of a lens obtained by decreasing to zero its central thickness.

Total track length The length from the vertex of the first surface of the first lens of a lens system to the image surface, along the optical axis.

Transverse magnification The ratio, $m = \dfrac{\overline{y}_o}{\overline{y}_i}$, of the image height to the object height.

Transverse ray aberration The transverse error $\vec{\varepsilon}$ of a ray at the image plane,

$$\vec{\varepsilon} = \frac{1}{n'u'} \vec{\nabla}_\rho W\left(\vec{H}, \vec{\rho}\right), \varepsilon_x = \frac{1}{n'u'} \frac{\partial W(\vec{H}, \vec{\rho})}{\partial \rho_x}, \varepsilon_y = \frac{1}{n'u'} \frac{\partial W(\vec{H}, \vec{\rho})}{\partial \rho_y}.$$

Triplet lens A lens system, or lens group, having three lens elements in proximity or in contact.

UV Ultraviolet radiation.

v-number A number that characterizes light dispersion by a material, given by $v = \dfrac{n_d - 1}{n_F - n_C}$.

Variator A movable, or changeable, lens group that changes the focal length of a lens system.

Varifocal A lens that provides a variable focal length within a range.

Veiling glare Non-image forming stray light that decreases the contrast of an image.

Vertex The point where the optical axis intersects an optical surface.

Vignetting The clipping, obstruction, of off-axis beams by an aperture, or ray limiting structure, other than the aperture stop.

Virtual image An image that cannot be cast on a screen because the image forming rays are diverging. The images formed by flat mirrors are virtual.

Wavefront A surface of equal optical path length, as measured from the object point. The surface of equal geometrical phase.

Working $F/\#$ See relative aperture.

Working distance The distance from the last physical structure of a lens to the image plane.

Zonal aberration Residual aberration that results from balancing different orders of aberration, or different aberrations.

Zoom lens A varifocal lens that maintains the image position stationary as the focal length changes.

Zoom ratio The ratio of the maximum focal length to the minimum focal length of a zoom lens. It is also known as zooming range.

Further Reading on Lens Design

Bentley, J., Olson, C. *Lens Design* (Bellingham, WA: SPIE Press, 2012).

Clark, A. D. "Zoom lenses," in *Monographs in Applied Optics*, Vol. 7 (London: J. H. Dallmeyer, Ltd., 1873).

Conrady, A. E. *Applied Optics and Optical Design*, Part I (New York: Dover, 1957).

Conrady, A. E. *Applied Optics and Optical Design*, Part II (New York: Dover, 1957).

Cox, A. *A System of Optical Design* (New York: Focal Press, 1964).

Dilworth, D. *Lens Design* (Bristol, UK: IOP Publishing, 2018).

Fischer, R., Tadic-Galeb, B., Yoder, P. *Optical System Design* (New York: McGraw-Hill, 2008).

Geary, J. M. *Introduction to Lens Design–With Practical Zemax Examples* (Richmond, VA: Willmann-Bell, 2002).

Gross, H. *Handbook of Optical Systems*, Vols. I–IV (Weiheim: Wiley-VCH, 2005).

Johnson, B. K. *Optical Design and Lens Computation* (London: The Hatton Press Ltd, 1948).

Kidger, M. *Fundamental Optical Design* (Bellingham, WA: SPIE Press, 2002).

Kidger, M. *Intermediate Optical Design* (Bellingham, WA: SPIE Press, 2004).

Kingslake, R. *A History of the Photographic Lens* (San Diego, CA: Academic Press, 1989).

Kingslake, R. *Optical System Design* (San Diego, CA: Academic Press, 1984).

Kingslake, R., Johnson, R. B. *Lens Design Fundamentals* (Amsterdam: Elsevier Inc., 2010).

Laikin, M. *Lens Design* (New York: Dekker, 2001).

Lummer, O. *Contributions to Photographic Optics* (London: MacMillan and Co., Limited, 1900).

Malacara, D., Malacara, Z. *Handbook of Lens Design* (San Diego, CA: Academic Press, 2013).

Mouroulis, P., Macdonald, J. *Geometrical Optics and Optical Design* (New York: Oxford Press, 1997).

Nakajima, H. *Optical Design Using Excel* (New York: Wiley, 2015).

Nussbaum, A. *Optical System Design* (Upper Saddle River, NJ: Prentice Hall, 1998).

O'Shea, D. *Elements of Modern Optical Design* (New York: Wiley, 1985).

O'Shea, D., Bentley, J. *Designing Optics Using CODEV* (Bellingham WA: SPIE Press, 2018).

Ray, S. F. *Applied Photographic Optics*, 2nd ed. (New York: Focal Press, 1997).

Riedl, M. *Optical Design, Fundamentals for Infrared Systems* (Bellingham, WA: SPIE Press, 2009).

Shannon, R. R. *The Art and Science of Optical Design* (Cambridge, MA: Cambridge University Press, 1997).

Slyusarev, G. G. *Aberration and Optical Design Theory* (Boca Raton, FL: CRC Press, 1984).

Smith, G. H. *Practical Computer-Aided Lens Design* (Richmond, VA: Willmann-Bell, 1998).

Smith, W. J. *Modern Lens Design* (Bellingham, WA: SPIE Press, 2008).

Sun, H. *Lens Design – A Practical Guide* (New York: CRC Press, 2017).

Taylor, H. D. *A System of Applied Optics* (London: Macmillan, 1906).

Velzel, C. *A Course in Lens Design* (Berlin: Springer, 2014).

von Rohr, M. *The Formation of Images in Optical Instruments* (London: H. M. Stationary Office, 1920).

Yabe, A. *Optimization in Lens Design* (Bellingham, WA: SPIE Press, 2018).

Index

Printed in the United States
by Baker & Taylor Publisher Services